MODELWORK

MODELWORK

The Material Culture of Making and Knowing

Martin Brückner
Sandy Isenstadt
Sarah Wasserman
Editors

UNIVERSITY OF MINNESOTA PRESS
Minneapolis
London

The University of Minnesota Press gratefully acknowledges the financial assistance provided for the publication of this book by the College of Arts and Sciences, the Department of Art History, and the Department of English at the University of Delaware.

Published by the University of Minnesota Press
111 Third Avenue South, Suite 290
Minneapolis, MN 55401-2520
http://www.upress.umn.edu

ISBN 978-1-5179-1089-1 (hc)
ISBN 978-1-5179-1090-7 (pb)
Library of Congress record available at https://lccn.loc.gov/2021025585.

Printed in the United States of America on acid-free paper

30 29 28 27 26 25 24 23 22 21 10 9 8 7 6 5 4 3 2 1

Contents

Part IV. Doing

Introduction
Modelwork

MARTIN BRÜCKNER AND SANDY ISENSTADT

This book brings together essays from a wide range of disciplines—including history of science and technology, music, literary studies, medicine, drawing and visual studies, art history, manufacturing, and more—to address questions of what models are and what they do. The answers are neither simple nor obvious, since, after all, everyone models: from artists and designers to prototype machinists and children. Because modeling is so pervasive, crossing professions and academic disciplines and permeating everyday life, it is difficult to define precisely: is it primarily an intellectual activity or a sensory, material one? A prospective task or a mimetic craft? A question or a solution? Where does one model end and another begin? From a humanities perspective, are accounts of the past or explanations of human behavior, for example, themselves generated by models of historical or social inquiry? When such models come to be internalized or taken for granted, do they in turn become models for subsequent investigation or action? In terms of understanding human culture, is it, to paraphrase an expression, just a matter of "models all the way down"?

How even to start talking about models? The *Oxford English Dictionary* dilates on the matter. Nearly fifty pages of single-column entries on the term's uses and etymological variants point in many and at times opposite directions, perpetually deferring any definitive or transcendent answer.[1] Discussions about models proliferate in the same way as those dealing with terms like *culture* or *nature*. Specialized meanings for models multiply, subdivide, and often drift between realms of measured professional discourse and casual consumer culture.

A good part of the difficulty in understanding models is that they are fundamentally hybrid. They amalgamate knowledge, both certain and indefinite, and practice, both critical and habitual. They are also bound to divergent modes of deliberation and mediation, whether theoretical foray or material craft. On one hand, historians of science, to single out one field, emphasize the role of models in advancing a conceptual overview of material phenomena. Though models help to reconcile theoretical understanding with observations of the physical universe, they are, in this view, mental constructs.[2] As philosopher of science Ian Hacking puts it, "a model in physics is something you hold in your head rather than your hands," or, as Ronald N. Giere, another philosopher of science, argues, models are most interesting when they are understood as "abstract objects" or "imaginary entities" that bypass the issue of their material manifestation or representation.[3] This position is common outside the history of science as well. The cultural theorist Jean Baudrillard, for example, argues that distinguishing traits—like color or size—of mass-produced consumer objects ultimately refer only to the underlying type, a nearly Platonic archetype that can exist only in one or another of its instantiations. Writ large, this means that for even the most generic objects to be distinctive, "it is essential that the model be no more than the idea of the model."[4] From this point of view, models are conceptual structures, and any material manifestation is no more than one expression at one moment of a more whole ideal.

On the other hand, in many cases, materiality matters and representation is deeply relevant, with "the idea of the model" powerfully conditioned by the making of the model and the ways in which it is perceived. Ball and wire models that sit on a tabletop, for instance, can represent molecules as easily as they can exemplify the motion of the planets around a star. Similar to behold, each model would nonetheless suggest patterns toward substantially distinct ends, bringing their physical traits, along with their material and scalar limitations, to bear on consequent definitions of substance, thereby subtly channeling subsequent theorization. For historians of science, as Soraya de Chadarevian and Nick Hopwood observe in *Models: The Third Dimension of Science* (2004), the authority of models derives from their cumulative modulation, because "models were made at all stages in processes of production. . . . Models used for teaching were often the same as those that guided research; models started as research tools and became teaching aids, but also vice versa."[5] Although not focused on models directly, scholars from Lorraine Daston and Peter Galison to Michael Lynch and Bruno Latour theorize the way physical representations, such as scaled objects, drawings, and maps, have the power to influence experimental outcomes.[6] Daston and Galison call these physical representations "working objects"; used in all sciences, such "working objects are not raw nature, they are not yet concepts, much less conjectures or theories; they are the materials from which concepts are formed and to which they are applied."[7] Empirical investigations exploiting such instruments are not innocent of the means used to pursue them. Latour's concept of "immutable mobiles," for example, posits nested representations that in their dissemination travel between venues—the lab, the archive, the journal article, and so forth—accruing with every circuit new shades of significance and explanatory power. They are abstractions that come to be materially freighted, like a palimpsest, with each incarnation.[8]

Latour's concept is especially valuable today in the face of proliferating digital representations that seem as intangible as ideas themselves. Computer models of worldly phenomena—from retail exhibits to climate patterns—conjure databases and algorithms rather than replicas or miniatures, dimensionless spaces rather than physical places. Regardless, they are representational devices that, in generating a display of some sort, create an "inscription," in Latour's terminology, a representation of a scientific object or, more generally, any phenomenon that is in turn inscribed upon by other instruments. As Latour suggests, such representations spring from mapping protocols that abstract selected bits of information, whether from the planets in their celestial orbits or the profits in a company's ledger, and put them into convenient forms, such as charts and tables. All this information, he says, is scaled so that it can be manipulated by scientists. Specifically, everything is distilled to two dimensions and put on paper: "at one point or another, they all take the shape of a flat surface of paper that can be archived, pinned on a wall and combined with others." This transcription of things by means of inscription is key to the regulation of nature and society alike.[9]

A model, too, is at some level a transcription of phenomena and a record of ways, including both analog and digital ones, to investigate those phenomena. When someone makes a model, whether a scientist or a programmer or an anthropologist, he or she makes strings of inscriptions, and it is in this context that models bear multiple traces of the intellectual and physical labor that went into making them.[10] Models are therefore records of the processes by which seemingly immaterial ideas are made manifest. But a model maker makes something else, something that follows from but also points beyond Latour's own model of the formation of scientific knowledge. Makers and users of models also engage with materials, negotiating material strengths and hardnesses, solid surfaces and invisible textures; they handle tools of various sorts that, as often as not, require varying degrees of dexterity and adjustment; sometimes, rather than hovering over charts and tables, model users and makers move around their models and physically engage with their distinct parts. All of these interactions affect the ways in which the model operates and is perceived. In thus guiding how something is represented, they also modify what is represented. That is to say, the model's materiality—its physical presence, its sensible materials, the volume it occupies—affects how we understand its referent. Even the mental models appreciated by Hacking and Giere will, as Latour notes, find their way at some point into some sort of material form and, as a result, be subject to physical and sociotechnical conditions, as well as resonate with the capabilities and limits of the human sensorium.

Yet, defining models in relation to materials and their attendant material culture is made difficult not only by the fact that there are so many forms of them but because they are usually distinguished in terms of function. Models may be largely descriptive, for example. Patent models, the subject of chapter 6 in this volume, would thus largely fall into this category as built explanations of, say, a machine of some sort. They are facsimiles and part of a larger system of representation that includes diagrams and explanatory text. Replicas—of a space shuttle or automobile, for instance—are also descriptive, as a tangible record of an object or a particular achievement or threshold reached. At the same time, descriptive models can also be prototypes or ideals, such as the form used to make a shoe or the epitome of a principled citizen. Another role for models is more propositional, a category that appears frequently both in the sciences and in design fields from architecture to video games. In this context, a model is like a sketch, a suggestion that is knowingly incomplete or even explicitly incorrect but whose inaccurate aspects are intended to spur refinement. It functions in such instances as a corridor of becoming, joining imagination with the realities of materials and natural forces like gravity and friction. Models used this way incrementally and interactively crystallize creative vision.

Models in general emerge from continuous acts of triangulations between an investigator of some sort and an object of reference or study. Semantic proxies for model, such as *description, facsimile, replica, sketch, epitome, prototype,* and even *vision,* all insinuate iterative material actions and interactions. In this way, a model is a mediating process, a mechanism for doing something. We call this process *modelwork.* With this term, we intend to incorporate an approach to the study of models that takes material factors into account, which would include relations of power and access along with physical conditions. The work of modelwork is not to assess how conveniently or accurately this or that model represents this or that thing or phenomenon. Rather, it encompasses the work of *working* the model, how the model is manipulated, what it "feels" like—to eyes and ears as well as to fingers—to work with it. Whereas the value of inscription, in Latour's discussion, is in making representations of micro- and macroscopic matters available to vision at the human scale, the notion of modelwork takes into account the physical factors that condition mental processes, including sensory inputs and modes of engagement. Even the most rigorous cerebral operations evince literary or poetic attributes that are drawn to a large extent from the physical world. As the philosopher John Sallis writes, however divorced from sensibility Immanuel Kant presumed reason to be (his own and the critical reason he critiqued), it is nonetheless—indeed, inevitably—bathed in metaphors that are themselves grounded in somatic experience. As Sallis writes, Kant's logical edifice—from its sturdy foundation to its towering conclusions, reached in stepwise fashion—entails an "architectonic metaphorics."[11] Likewise, material culture pervades models of even the most abstract phenomena or the most distant memories, making its traces felt in every step of modelwork.

The essays gathered here under the conceptual umbrella of modelwork focus less on what constitutes a model than, appropriately, on the work that models do. They presume that even descriptive models perform work of some sort; they *do* something beyond mere reference or representation. Despite changes in technology, modeling in any field always involves, or, to sharpen the point, comprises, its own unique kind of labor. This is not the kind of work that Raymond Williams associates with toil and paid employment.[12] Nor is it the "immaterial labor" of cultural work proposed by Marxist theorists like Michael Hardt and Antonio Negri, who examine work in a rapidly transforming "cognitive capitalism," where the production of immaterial and symbolic goods eclipses other sorts of productive activity.[13] Modelwork occupies a distinct place in such economies of labor because it straddles fault lines of material and immaterial realms. It makes sense of a materially present object by reaching into the unseen, the not-yet, or even the long-gone. It presumes that models are in general contrivances with which people reason and learn and, like any tool, simultaneously facilitate and constrain actions. While models, once made, can appear self-evident and thereby naturalize social formations and obscure underlying ideological motivations, the task of modelwork, as a paradigm for historical and theoretical investigation, centers on the boundaries and definitions that aim to stabilize and demonstrate otherwise ungraspable or ambiguous ideas with the air of self-evidence.[14]

The idea of modelwork expands in several ways on Latour's discussion of "paperwork" as a way to capture a dual modality, a toggling back and forth between imagination and material-based production.[15] According to Latour, paperwork considers the crafts of writing, reading, making images, and visualization in the pursuit of scientific knowledge, rather than building an account on either a strictly materialist or ideational foundation. All the scientific experiments we know of, he points out, including their raw data, instruments, venues, documentation, and so on, have been committed to paper, appearing as a kind of literature comprising words, charts and tables, and images. This is the two-dimensional world of inscriptions mentioned earlier, representations of events that are both stable and portable and that, as a result, allow—invite, even—recombination at different scales, with allied materials, in new contexts and for new purposes.[16]

The essays in *Modelwork* recognize both the layers and the structures of inscription that accrue as ideas pass through embodied and ideational phases. Practices both material and immaterial leave their traces in and on models, creating something like a representational residue. In turn, this patina can transform our understanding of the model, attesting to the subjective assimilation of even the most objective matters. But these essays also develop the idea further by acknowledging the unique role of models in human inquiry and creativity and by taking account of specific materials and material substitutions, changes in scale, and, as a result, positional and spatial conditions. They show that modelwork—the dialogue of model, model maker, and model user—cannot be reduced to two-dimensional drawings and ledger entries, letters and textbook excerpts, or questions solely of knowledge. Modelwork points toward the multiple labors of modeling, including the maker's initial process of translating ideas across media, that is, energizing a circuit passing through mind, paper, and model, as well as subsequent iterations where the physical model, be it amplified or minimized with specific attributes, informs the understanding of the model. In short, as well as helping us think, models are apprehended; they are made; they do things.

To the extent that modelwork expresses a material culture perspective on "paperwork," we remain within the same intellectual traditions examined by Latour, largely Western in outlook and addressing the last three centuries. Assessing the value of modelwork as a new critical approach meant retaining other parameters common to these traditions, including the interplay of economic paradigms, institutional and professional roles, and technological practices, such as inscription protocols. Our aim is to build on a tradition of inquiry rather than apply existing approaches to new conditions like globalization or bring new approaches, such as "blue humanities," to existing conditions. Moreover, the very idea of *model* is especially representative within that tradition. The word descends from the Latin term *modulus,* a diminutive of *modus,* meaning "measure" or "extent," especially a proper or just measure. The word permeates European languages in a remarkably stable form and meaning, and whether *model* (En.), *modèle* (Fr.), *Modell* (Ger.), модель (Rus.), or *modelo* (Sp.), it encompasses a penumbra of related meanings, including manner, mold, example, a sitter for an artist, a template, and, adjectivally, perfect, as in a model citizen. Likewise, it is expansive, effectively appearing across all domains, from the arts and humanities to the social, natural, and applied sciences.

In each translation, models lay at least some claim to a dispositive relationship between mind and matter, if not a complete universality of meaning; by contrast, the inherent material culture perspective behind modelwork emphasizes issues of translatability and generative transformation when ideas traverse material modes and material configurations condition ideation. Similar to other material culture approaches, modelwork occupies an intermediary platform that encourages critical engagement both with models and with the variable processes that they shape.[17] It insists on an interplay of mind and matter, a biography and social of life of things, and a recognition of the power to script and control personal and societal actions.[18]

Consequently, the volume is divided into four sections—"Knowing," "Sensing," "Making," and "Doing"—each mapping out particular registers of epistemology, phenomenology, facture, and action even as they overlap and reconfirm the intricate ways in which such heuristic principles interlace. Taken together, the range of perspectives that inform *Modelwork* creates its own model of modeling as a transdisciplinary and profoundly humanistic pursuit. The essays demonstrate that the interchange between the imaginative and the material occurs as much in the humanities and liberal arts as it does in science and technology. Given the ubiquity of models, this volume also points toward larger conceptual debates about the way in which models of the past as well as new digital ones profoundly shape the world around us. By providing richly layered insights into our current moment, in which accelerated speeds and expanded scales are troubling familiar modes of sense perception, notions of knowledge, forms of work, and embodied practices, *Modelwork* offers new exploratory tools for understanding the generative powers of modeling.

KNOWING, SENSING, MAKING, DOING

"Knowing," the first part of *Modelwork,* considers epistemological aspects of modeling. Tapping philosophical debates about knowledge production in the sciences and the humanities, the essays in this section explore the ways in which modeling is both a constitutive means of understanding complex systems and a challenge to assumptions about those systems. To create a model is to presume a degree of authority for that model that is at the same time subverted by the translation of one thing into different scales, different materials, and different modes of apprehension. Modeling creates a kind of shadow discipline for any academic specialty. Researchers make use of methods and forms internal to a field but do so by means of a set of skills distinct from those that practitioners usually consider central to their discipline. Rarely a subject in themselves, models are a means by which to investigate other subjects. By virtue of their extradisciplinary nature, they can create a fresh perspective on a field of study, complete with framing elements, foreground, background, and so on, even a position for the observer, which lends intentionality to the questions being asked of any particular model. The process can stimulate new insights on a subject distinct from other methods that can create conceptual distance. The transformative potential of models might well reside in their being simultaneously analogous to the phenomenon being investigated and foreign to the discipline in question.

In this sense, the model is itself an epistemological analog. Testing the logic of the model's authority, Annabel Jane Wharton reflects on philosopher Nelson

Goodman's characterization of models as "promiscuous" to define the model at a time when, for many, the model's persuasiveness is presumed to rest on its autonomy. She concludes with a definition that characterizes modeling without at the same time containing it. Peter Galison looks at the role material models can play as mechanisms that mediate different forms of scientific reasoning, especially in regard to immaterial systems, such as electromagnetic fields, economic relations, and black holes. Rather than being an ersatz or substandard theoretical mode, physical models can combine with other approaches to generate a deep and engaging understanding apart from description, explanation, or prediction.

The second part, "Sensing," explores the phenomenological dimensions of modeling. Models have been used to illustrate the work of the senses and have, over time, changed our conception of the human sensorium altogether, especially as it has expanded to include contexts of cultural and environmental perception. Three essays track some of these developments, exploring how modeling emerges from and reflexively revises the act of sensing. Models, once made, insert themselves between a human viewer or agent and the phenomenon to which the model refers. The model may not only delimit the scope of observation but can to some degree determine what is *observable*. Contributors to this section consider how the optical, aural, and tactile properties of models inflect and are reciprocally regulated by subsequent acts of cognition, imagination, and making in the dynamic cycle of modeling. Hilary Bryon looks at the way in which the "isometrical perspective," a drawing system developed in the late eighteenth century, advanced a critique of linear perspective and multiplanar orthographic projections by combining the pictorial aspects of the former with the measurable scale of the latter. Initially used for material models of mechanical principles, the system modeled at once analogic and iconic perspectives on the physical world. Also examining the sense of vision, Christopher J. Lukasik explores the rise of drawing books, tracing the shift in their goals from training the viewer's visual imagination to an ever more accurate manual transcription of the visual world and, ultimately, a model of modern communication. By contrast, Martin Scherzinger investigates the modeling of the musical ear in the present day by considering the kinds and forms of data that are recognized by current music software and their relations with various musical practices. Demonstrating how music is captured by extramusical concerns, from instrument simulation technologies to Western postindustrial contexts, he explains how musical productions fusing historical and social mediations with long-standing human neural responses create a new technosensory form of musical temporality.

Part III, "Making," considers the making of a model as a combinatory and formative process that binds physical elements with abstract forms of production.

The essays pose basic questions: How are models made? How do the conventions and practices of making affect cognitive or affective aspects of modeling, and at the same time, how does the modeling process—conceived as a kind of dialogue between physical and conceptual elements—revise and reframe what is being made? To put it another way, if model-making is both a mimetic and a prospective process, what, exactly, gets made—beyond the physical artifact—when we model? A major function of many models is to map a converging trajectory between imagination and embodiment. In this propositional mode, models invite an imaginative inhabitation of structures or situations that are more ambitious than the model at hand.

Examining the archive of patent models, Reed Gochberg discloses the complex relationship between model-making, aesthetics, and economic considerations. Because patent models exceeded bureaucratic and legal functions, they frequently became a type of proving ground for middle-class consumers who saw underpinning any one particular mechanism a human future steeped in material inventions. Two plaster sculptures displayed at the 1893 Columbian Exposition in Chicago are the focus of Catherine Newman Howe's study of artists and designers who tried to craft a statistically average and, by extension, ideal American body. Taking into account thousands of composite photographic portraits and averaged measurements, they proposed precise models of perfect physical fitness. Leaving the realm of analog modeling, Seher Erdoğan Ford introduces the "hypermodel" as a digital re-creation not only of a lost artifact but also of the incomplete, ambiguous, and often conflicting history of efforts to recover the loss. She offers a case study of the ruins of a fifth-century Byzantine basilica in present-day Istanbul to make manifest the active work of computational modeling within the act of historical interpretation and scholarly reconstruction.

The essays in Part IV, "Doing," envision models as agents that act upon both their environments and the people who make and use them. The authors point to varying degrees of separation that models can attain from the things that they refer to. While models are, almost by definition, simpler than the things to which they refer, they are nonetheless suggestive in that they can shift attention from an absent real or ideal referent to a manifest and immediate notation. At the same time that the model draws into presence selected continuities with its referent, it also eclipses differences. The model becomes an embodied assertion of self-presence, which might well be a foundational attribute of models: they sit between us and the world we want to build or understand or remember. As such, the model may elicit responses and questions that have as much or more to do with its own internal logic as with the cultural or structural logic of whatever phenomenon it is modeling.

Turning to nineteenth-century "dissected plates" and the intricately layered flaps published in anatomical books, Juliet S. Sperling recovers procedural models that invited readers to pinch and manipulate paper tabs that rehearsed the complex manual maneuvers of surgery and to trace with their own hands the changing contours of a woman's uterus as it swells over the course of gestation. Industrial procedures are the subject of Lisa Gitelman's look at the history of footwear fabrication, centered on "shoe lasts," the wooden model on which shoes were made. She argues that modern shoe production was an arena in which the physicality of models loomed large, as the industry not only depended on models of shoes but became itself a model of factory-based production in which labor, industry, and locality were reimagined. Finally, attending to nonrepresentational approaches developed by digital databases and interactive platforms engaging word, text, and literature, Johanna Drucker further shifts our understanding of modeling from the product—such as a working plan, a data model, a design, even a set of procedures, as in modeling behavior—to the process itself, from a representation or secondary presentation of prior notions to a primary, constitutive activity.

Throughout the volume, the various models discussed all pose the question, Can you imagine that . . . ? Because they reference something outside themselves as well as their ineluctable incompleteness, models invite a creative leap. They inhabit the subjunctive. As such, they hold great potential for change, whether that change be negligible or decisive. They are instruments of power insofar as they help us strengthen our intellectual grasp of the world or rehearse a possible course of action or generate a new object or, all too often, rationalize the oppression of peoples. In this way, models are as much a part of the infrastructure of modern social life as mass production, perpetual change (or creative destruction), nation-states, racial hierarchies, and consumer culture. They appear in one form or another in virtually everything from cosmology science to child's play. Indeed, the desire to grasp the workings of the world and even the universe might well descend from childhood memories of models—building blocks, dollhouses, or train sets—that sponsored a sense of control over an environment. In other words, among the many other things they do, models model mastery. In miniaturizing and simplifying an aspect of the world, models offer provisional access to that world (or hold it in abeyance) and introduce the idea of reordering it toward some end.

Such access, however, comes at a cost. The value of modeling for remaking the world is matched only by the potential for misunderstanding it. A model's operational continuities with its referent are made possible only by creating discontinuities, whether in scale, material, or degree of force—or by discounting complexities, such as having incomplete or even false information or any of the usual infelicities of human judgment. To the extent that a model offers insights, these discontinuities can be disregarded or even forgotten. As a result, the constitution of the model might be mistaken for the thing being modeled, the map taken for the territory, so to speak. As new representational tools advance explosively in our own digital age, along with new modes of inquiry, the uses and ubiquity of modelwork across fields demand that we reflect on the ways in which models inevitably both entail and veil interpretation and world-making.

NOTES

1 To be more precise, in the online *Oxford English Dictionary,* the term registers with more than 600 definitions, more than 450 etymological variants, and approximately 3,500 quotations. See *OED,* s.v. "model." The first essay in this volume, by Annabel Wharton, offers perhaps the best and fullest account of modeling, a precise definition that takes some pages to arrive at.

2 Mary S. Morgan and Steve Woolgar, *Models as Mediators: Perspectives on Natural and Social Sciences* (Cambridge: Cambridge University Press, 1993).

3 Ian Hacking, *Representing and Intervening: Introductory Topics in the Philosophy of Natural Science* (Cambridge: Cambridge University Press, 1983), 216; Ronald N. Giere, *Science without Laws* (Chicago: University of Chicago Press, 1999), 5.

4 Jean Baudrillard, *The System of Objects,* trans. James Benedict (London: Verso, 1996), 155. In this example, Baudrillard is speaking of automobiles.

5 Soraya de Chadarevian and Nick Hopwood, *Models: The Third Dimension of Science* (Stanford, Calif.: Stanford University Press, 2004), 3. Studying empirical "hard" models, the essays in this volume emphasize the recovery of models and their expository role in collections, research, and teaching; they are less concerned with abstract definitions of what constitutes a model or the process of modeling.

6 See Bruno Latour, "Visualisation and Cognition: Drawing Things Together," *Knowledge and Society: Studies in the Sociology of Cultures Past and Present* 6 (1986): 1–40, and Latour, *Science in Action: How to Follow Scientists and Engineers through Society* (Cambridge, Mass.: Harvard University Press, 1987); Michael Lynch and Steve Wolgar, eds., *Representation in Scientific Practice* (Cambridge, Mass.: MIT Press, 1990); Peter Galison, *Image and Logic: A Material Culture of Microphysics* (Chicago: University of Chicago Press, 1997); or Lorraine Daston and Peter Galison, *Objectivity* (Cambridge, Mass.: Zone Books/MIT Press, 2007).

7 Lorraine Daston and Peter Galison, "The Image of Objectivity," *Representations* 40 (1992): 85.

8 Latour, *Science in Action,* 227.

9 Latour, 228–29.

10 Outcome assessments, especially in education, offer a glimpse of the way inscriptions are generated. See Jo Rycroft-Malone and Tracey Bicknall, eds., *Models and*

Frameworks for Implementing Evidence-Based Practice: Linking Evidence to Action (London: Wiley-Blackwell, 2010). For outcome assessments at the intersection of mathematical, computational, and concrete modeling, see Michael Weisberg, *Simulation and Similarity: Using Models to Understand the World* (New York: Oxford University Press, 2013).

11 John Sallis, *Spacings—of Reason and Imagination in Texts of Kant, Fichte, Hegel* (Chicago: University of Chicago Press, 1987), 21–22. On the presence of materiality in language, see George Lakoff and Mark Johnson, *Metaphors We Live By* (Chicago: University of Chicago Press, 2003), esp. chapters 8, 18, and 19.

12 See the entry "Work" in Raymond Williams, *Keywords: A Vocabulary of Culture and Society* (New York: Oxford University Press, 1976), 282.

13 See Michael Hardt and Antonio Negri, *Multitude: War and Democracy in the Age of Empire* (London: Penguin, 2005).

14 Addressing novels, Philip Fisher characterizes cultural work in a way that also applies to models more generally as being "work that, once done, becomes obvious and unrecoverable because it has become part of the habit structure of everyday perception." Fisher, *Hard Facts: Setting and Form in the American Novel* (New York: Oxford University Press, 1985), 6.

15 Latour, "Visualisation and Cognition," esp. 25–28. While our understanding of modelwork resonates with design theory and the critical lens to look for a model's "affordances"—after all, its functions "refer to perceived and actual properties," and its iterative forms also "describe the potential uses or actions latent in materials and designs"—this approach does not capture the way in which models vacillate between being abstract configurations of a process and the sense that whatever material configuration a model assumes, it is at once decisive and yet arbitrary. See Donald A. Norman, *The Design of Everyday Things* (New York: Basic Books, 1990), 9, and Caroline Levine, *Forms: Whole, Rhythm, Hierarchy, Network* (Princeton, N.J.: Princeton University Press, 2015), 6. See also Williams, *Keywords*, 259.

16 As if wary of the two-dimensionality of graphics and other representational modes of inscription, Latour proposes that "information, traces, goods, plans, formats, templates, linkages" are movements rather than shapes inhabiting flattened space. See his *Reassembling the Social: An Introduction to Actor-Network-Theory* (New York: Oxford University Press, 2005), 204–5.

17 On material culture studies and its approaches inhabiting an interstitial place in the humanities, see Victor Buchli, introduction to *The Material Culture Reader*, ed. Victor Buchli (New York: Berg, 2002), 15.

18 Buchli calls modern material culture a "super-category" coming into focus during the Great Exhibition of 1851 with the display of newly made consumer goods (models) and their commercial displays (a version of modelwork); see Buchli, 3, 5. On approaches to material culture that also represent Western ideology, see on

knowledge production Michel Foucault, *The Order of Things* (London: Routledge Press, 1970), or Latour's *Science in Action*; on mind in matter, see Jules David Prown, *Art as Evidence: Writings on Art and Material Culture* (New Haven, Conn.: Yale University Press, 2001); on the biography of material cultures, see Arjun Appadurai, ed., *The Social Life of Things* (Cambridge: Cambridge University Press, 1986), and select object lessons in Fiona Candlin and Rayford Guins, eds., *The Object Reader* (New York: Routledge, 2009); on the power of material culture's patina scripting human behavior, see Daniel Miller, *The Comfort of Things* (New York: Polity, 2009), or Robin Bernstein, *Racial Innocence: Performing American Childhood from Slavery to Civil Rights* (New York: New York University Press, 2011).

Part I

Knowing

1

Defining Models

ANNABEL JANE WHARTON

PROMISCUITY

In *Languages of Art,* the distinguished American philosopher Nelson Goodman alleges the model's profligacy and even suggests its expendability: "Few terms are used in popular and scientific discourse more promiscuously than 'model.' A model is something to be admired or emulated, a pattern, a case in point, a type, a prototype, a specimen, a mock-up, a mathematical description—almost anything from a naked blonde to a quadratic equation—and may bear to what it models almost any relation of symbolization. . . . 'Model' might well be dispensed with in all these cases in favor of less ambiguous and more informative terms."[1] Here Goodman articulates the broadly held understanding of "model" as a generic term naming a vast array of very different things: it is attached to too many kinds of things to be very helpful. The *Oxford English Dictionary*'s entry on *model* is nearly nine columns in length. The *OED* lists nearly two dozen meanings of *model,* not including compounds, technical terms, and obsolete uses. Certainly there are a large number of model types—from scientifically engineered mice and naked blondes through toys and architectural prototypes to economic pie charts and climate change models. All of these model types obviously take very different forms—organic, figural, computational, and so on. And, of course, not only are their ontologies distinct but the ways in which they act—paradigmatically, ludically, diagnostically, and so on—are also distinct. Models are indeed a prolific class of things.

Models' fecundity provokes Goodman to slander. Though he charges language, not models, with promiscuity, he obviously holds models responsible for discourse's shameless behavior. After all, promiscuity implies careless, indiscriminate acts of libidinous wantonness of the sort elicited by those he calls "naked blonds." *Modelwork* documents the sorts of provocative acts of models that incited Goodman's abuse. But although Goodman's frustration with models is by no means unfounded, I question the legitimacy of his final verdict. Goodman's condemnation of the word *model*—that it "might well be dispensed with"—is not only unenforceable but also unjust. Despite being denotationally libertine, *model* is not, as a descriptor, utterly profligate. To help discipline the term, I offer here a short but inclusive definition for "model."[2] The libidinous acts of models can be better understood as generative once their common properties, their community, are identified.

COMMUNITY

To begin with the obvious: all models have referents; every model stands in some relationship to that to which it refers. Another shared feature of models: not only must they have a referent but they must also differ in some way from that referent. A model is not a model of an object if it is identical with it—rather, it would then be a clone, a simulacrum, a double. The ways in which a model may differ from its referent are at least as varied as its referent and the means by which the model relates to its referent. Models conventionally differ from their referents by changes in their scale (usually, but not always, smaller), their complexity (usually, but not always, simpler), and/or their material (usually, but not always, cheaper). However, models also differ ontologically from their referents. Is a model's way of being in the world ever the same as its referent? Does it ever serve the same function, elicit the same forms of attention, or come into existence in the same manner?

Goodman complains, quite rightly, that models can model just about anything, real or imaginary, material or conceptual—buildings and behaviors, theories and threats, weather patterns and monetary flows, qualitative values and quantitative data. Models not only have a vast range of referents but they also have a vast range of ways of relating to those referents. As Goodman recognizes, a model "may bear to what it models almost any relation of symbolization." The rich array of ways by which a model might connect with its referent—that "relation of symbolization"—is the model's semiotics. To suggest something of the variety of these relations, I deploy Charles Peirce's familiar observations on the means by which signs signify. His terminology offers a convenient shorthand for the wide spectrum of linkages between a model and its referent—icon, index, and symbol.[3] Peirce names these expressions of the sign according to how each establishes its relation with its object (Figure 1.1).

An icon (εἰκών), for Peirce as well as for the ancient Greeks, is an image: it conveys its meaning through visual resemblance. Scale models, like a souvenir Eiffel Tower or the Dome of the Rock in the video game *Assassin's Creed,* relate to their referents *iconically*; they look like their archetypes. A symbol is an abstraction that depends entirely on convention and social habit to convey its meaning. An equation or an algorithm is an alphanumeric construction that stands in for a

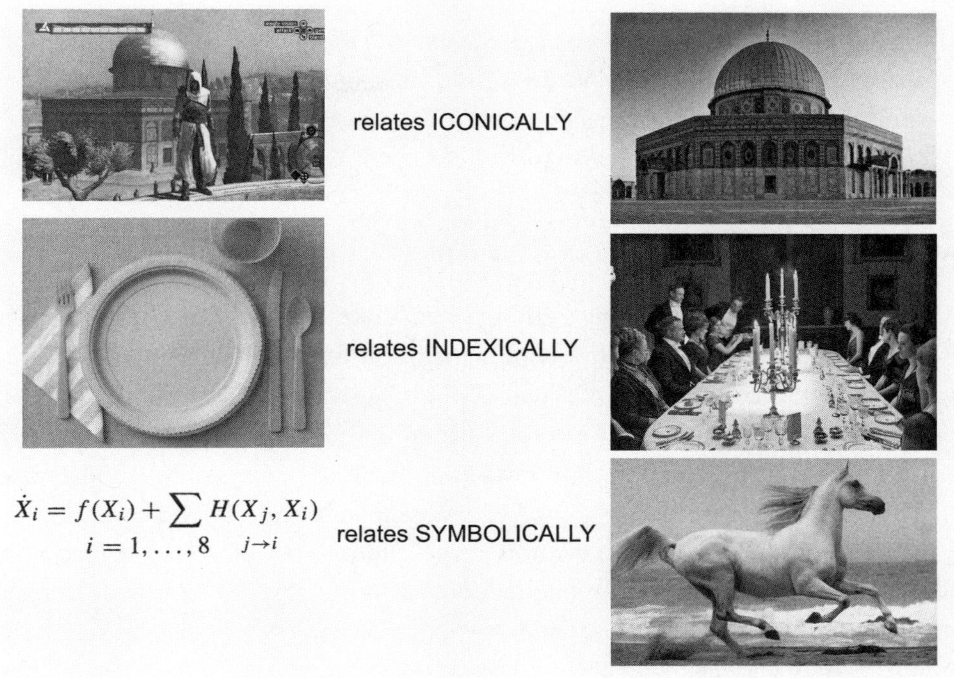

FIGURE 1.1. Semiotics. Equation from Pietro-Luciano Buono, "Models of Central Pattern Generators for Quadruped Locomotion: Secondary Gaits," *Journal of Mathematical Biology* 42 (2001): 327–46. Chart constructed by the author.

narrative proposition, theory, or phenomenon.[4] Index, which stands in an ambiguous, secondary relationship with its object, is more complicated than either the icon or the symbol.[5] For Peirce, an index draws attention to an object through some physical contiguity. The conventional example of Peirce's index is smoke, which is the indexical sign of a fire. It conveys its meaning through the traces it leaves of its object. For Peirce, both the object and its sign are material and singular: the material form of the index implies both the uniqueness of its object and its own independent existence. But the range of the objects to which an index relates is much broader in common language than in Peircian theory.[6] The *Oxford English Dictionary* defines an index as "that which serves to direct or point to a particular fact or conclusion." A market index is the aggregate value of a selection of stocks or other investment vehicles used as a sign of the economy. The twisted spire of St. Mary's in Chesterfield provides an index by which the town's inhabitants locate themselves. A host's arrangement of place settings for a dinner party denotes his familiarity with the cultural habits of his class; it is an index of his location in the social hierarchy. From this discussion, the tentative first sentence of my definition of models may be offered:

A model is a thing that has a referent (material, ideal, conceptual, or imaginary) to which it semiotically averts (iconically, indexically, or symbolically) but from which it differs in significant ways (in complexity, scale, material, function, ontology, and the like).

AGENCY

My identification of a model as a "thing" is, of course, untenable. "Thing" is too material, as we have been told by phenomenologists from Heidegger to Bill Brown, who have bludgeoned us with it.[7] "Thing" doesn't readily admit ideas, like utopias, or graphic abstractions, like algorithms. "Thing" is also inanimate. It doesn't readily admit sentient beings, like the supermodels on fashion runways or the model mice in a scientist's laboratory. For "thing," I propose to substitute "agent."[8]

The agent of my definition is not that agent found in philosophy or common discourse—that is, an actor embedded in heavily freighted concepts of human morality, personal intentionality, and free will. Rather, it's the agent of chemistry or business. In chemistry, an agent is merely a substance, but a substance that has physical, chemical, or medicinal effects on proximate things. In business, an agent is "a person [the agent] employed by another [the principal] to act for him."[9] If, in chemistry, consciousness and intentionality are extraneous to agency, in business, agency also loses most of its ethical baggage. In treating *models* as agents, I am, as in chemistry, naming entities—whether spatial objects, abstract notations, sets of thought, or even human beings (like drunk drivers)—that have an effect on their physical or mental environments without requiring of them any inkling of consciousness or intentionality. A model-agent differs from a chemical-agent in that its effects on its setting are much less predictable. The model-agent is also like a business-agent. It may be assigned tasks by its principal (maker, patron, or owner). But if and how those tasks are carried out is not contractually binding. That is, however a principal might intend a model to act, she can never fully control its actual performance in the world.

I am by no means the first to suggest that models are agents. The distinguished economist Margaret Morrison effectively demonstrates models' agency, identifying models not just as agents but as "autonomous agents."[10] Models are, she claims, independent both of the world and of the particular theory about the world that they are designed to demonstrate. She convincingly argues that if models are going to act effectively, they must be distinct from the theories that inspired them and from the empirical data that they are attempting to more fully understand, explain, or reveal. I would push her distinction further, adding that models, once

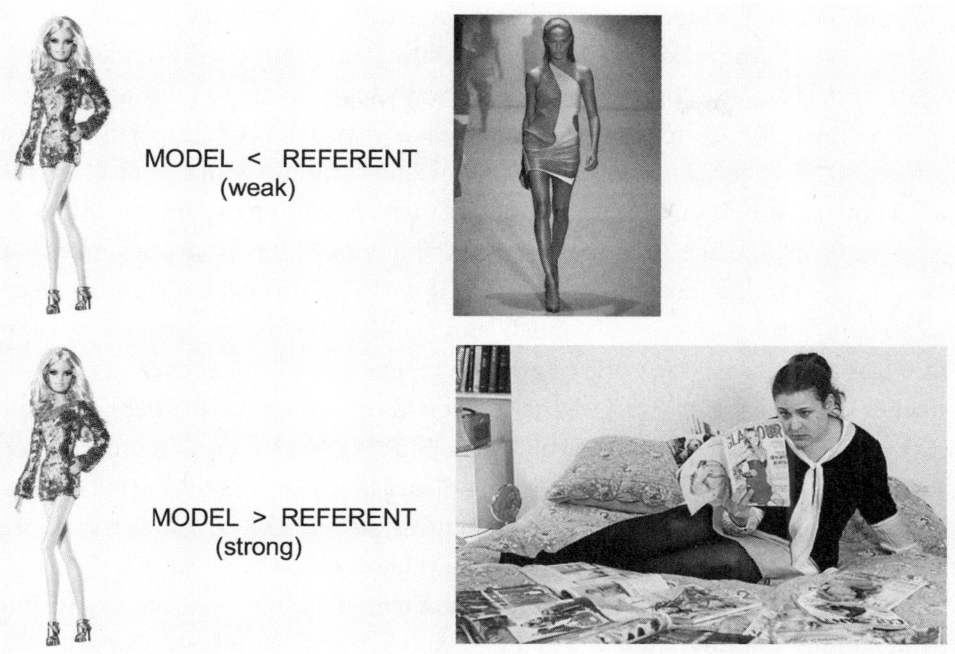

FIGURE 1.2. Weak and strong models. Equations and chart constructed by the author.

complete, are also independent of their makers and their consumers—like children are independent of their parents or, perhaps better, like buildings are independent of their architects. The agent in my definition is, then, an actor that is not only without consciousness or intentionality but also one that is independent—at least insofar as anything in the world is independent.

Independence contributes to agency, but it does not assure its liveliness or its efficacy. A model-agent, like any other agent—human, chemical, natural—can be strong or weak. That strength or weakness depends on its own integrity, its particular social location, its luck. A model-agent is peculiar, however, because its weakness or strength is further conditioned by its referent. Most obviously, the authority of a model may depend on the authority of its archetype. A strong referent certainly does not ensure the power of its model, but it may well contribute to it. The aura of the Dome of the Rock clings to its model in *Assassin's Creed.* The Yale investment model was adopted by other universities, often disastrously, in part because Yale used it.[11]

The weakness or strength of the model-agent is not only affected by the status of its referent. It is more peculiarly determined by the conditions of their relationship. A *strong* model acts as a dominant subject that determines its weak object. An architect's model, for example, may be expected to work as an archetype for the building to be constructed. The market may change to conform to a rhetorically powerful economic model made of it. Joan Smalls, supermodel, makes clothes look fabulous, though she makes women feel ungainly. There are also *weak* model-agents. Weak models act like copies. Copies are always subordinate to their archetypes. However much fun it is to play with, a toy car is a weak model of the real thing. This distinction between weak and strong has much in common with Clifford Geertz's well-known distinction between "models of" (the rendering of complex things in "synoptic form" to make them apprehensible) and "models for" (synoptic renderings allowing the production of more complex things).[12] But my distinction between "weak" and "strong" is, critically, much less clear than that made between "of" and "for" by Geertz, who was not concerned with agency. Model-agents, in contrast to other kinds of agents, exercise that agency in part by oscillating between their weak and strong potentials. A Barbie doll, which is a weak copy of a supermodel, may well contribute directly to the eating disorder of her possessor. This oscillation of the model in relation to its referent gives the model its important capacity to make its observers rethink the archetype (Figure 1.2).

Like models themselves, interpretations of a model can be weak or strong. But to offer a valid interpretation of a model, a commentator must be conversant with the rules of the model-type on which she remarks. Model-types all have their own particular conventions. Physics equations, economic graphs, and even axonometric drawings may well not be accessible to those unfamiliar with their symbolic codes. That exegetes must understand the etiquette of the models on which they comment is obviously the case with computational forms. But it is equally so with less abstract examples. A nuanced reading of David Levinthal's ironic images of war, published in *Hitler Moves East: A Graphic Chronicle, 1941–1943*, requires its interpreter to recognize that the figures are model toys and to know that model toys are ordinarily ludic objects (Figure 1.3).

This discussion of a model's agency and hermeneutics sanctions the modification of the first sentence of my definition and additions to it:

> A model is *an autonomous agent* that has a referent (material, ideal, conceptual, or imaginary) to which it semiotically averts (iconically, indexically, or symbolically) but from which it differs in significant ways (in complexity, scale, material, function, ontology, and the like). *A model assumes its interpreter's familiarity with its particular hermeneutic conventions. It can be strong or weak.*

FIGURE 1.3. David Levinthal, photograph of model toys from *Hitler Moves East,* 1975. Photograph courtesy of David Levinthal and with his permission.

OPERATION

Models obviously work differently in different disciplines and for different audiences. They may also act in a variety of ways over time for the same observer. Not only are their specific functions as varied as their makers and consumers but also, as autonomous agents, they often act in ways unintended by either. Because models always either exceed or resist the work assigned to them, attempting to enumerate their particular modes of labor in a definition is pointless. Nevertheless, the attempt to formulate what makes models into a community discloses something of the collective ethics of models' operations. I use ethics here to mean simply "the rules of conduct recognized in respect to a particular class of human actions or a particular group, culture, etc."[13] Models exhibit a certain shared standard of acceptable behavior in how they act if not in what they are meant to accomplish. Good models work to make the world in some way or another more accessible. In the model-making process itself, in the action of modifying the model to get it to act more like its referent, or even in the deep engagement of critically thinking about a finished model, the model offers a fuller understanding of the bit of mental or material existence that it models.

The epistemic work of model-making and manipulation is broadly acknowledged. Certainly the making of a model is speculatively generative. Sketching the Buddhas of Bamiyan or writing a book about models, the artist or author usually learns more about her object than does the viewer of the drawing or the reader of the book. The active scrutiny of a finished model, the critical search for its weaknesses and strengths undertaken before accepting its interpretation of the world, is also a kind of manipulation that generates knowledge. In *The World in the Model,* Morgan observes, "Modeling is not an easy way to find truths about the economy, but rather a practical form of reasoning for economists, a method of exploration, of enquiry, into both their ideas and their world."[14] As Baden-Fuller and Morgan observe, "we do not learn much from looking at a model—we learn more from building the model and from manipulating it."[15] The distinguished urban planner Marcial Echenique similarly suggests that "the main purpose of a model is to provide a simplified and intelligible picture of reality in order to understand it better. It should be possible to manipulate the model in order to propose improvements in [the model's representation of?] reality."[16] Augmenting a climate change model with interactive atmospheric aerosols, atmospheric chemistries, and representations of the carbon cycle not only provides a more accurate image of the world but also allows the modeler new insight into the world's processes. The child playing with a doll's house adds furniture to it or rearranges its appurtenances to make it correspond more closely to lived or imagined conditions. In that process, he better understands the domestic relations of objects and their possessors.

The doll's house, like other model toys, suggests something further about the way that all models work. "Play," at least if it is understood in terms of its definition by the Dutch historian Johan Huizinga, is a productive way to understand models and their manipulation.[17] In *Homo Ludens,* Huizinga argues that play is necessary, free, disinterested, reproducible, space- and rule-bound, and even magical.[18] Some aspects of play articulated by Huizinga would seem usefully applicable not only to the manipulation of models but also to models themselves:

> Inside the play-ground [model-space] an absolute and peculiar order reigns. Here we come across another, very positive feature of play [the model]: it creates order, is order. Into an imperfect world and into the confusion of life it brings a temporary, a limited perfection. Play [the model] demands order absolute and supreme. . . . Play [the model] casts a spell over us; it is "enchanting," "captivating."[19]

The pleasures and dangers of models may sometimes be more apparent in nonscientific models than in scientific ones.

The proposition that models are vehicles for investigating the world is broadly assumed by those who use them. Models provide a mechanism for probing the world's reality. Karl Popper, the distinguished philosopher of science, goes further, arguing that models are a means of revealing the truth of the world. In a paper presented to economists at Harvard, he offers his definition of truth: "There is an old answer to the question: 'When is an assertion, or a proposition, or a statement, or a theory, or a belief true?' The answer is: An assertion is true if it corresponds to, or agrees with, the facts."[20] Popper further defines the work of models and theories:

> We have no objection to the assertion that all scientific [models and] theories are instruments—either actual or potential instruments. But we assert that they are not *merely* instruments. For we assert that we may learn from science something about the structure of our world: that scientific [models and] theories can offer genuinely satisfying explanations that can be understood and so add to our understanding of the world. And we assert—this is the crucially important point—that science aims at truth, or at getting nearer to the truth, however difficult it may be to approach truth, even with very moderate success.[21]

Popper does not specify the jobs done by scientific and social scientific models and theories but forcefully argues that they serve as vehicles for moving their makers and manipulators closer to the truth. Popper lends his pre-postmodern position credibility by acknowledging that models and theories never reach their goal. To the question "Can a model be true?" Popper answers,

> I do not think so. Any model, whether in physics or in the social sciences, must be an over-simplification. It must omit much, and it must overemphasize much. . . . It seems to be quite unavoidable in the construction of models, both in the natural and in the social sciences, that they over-simplify the facts, and thus do not represent the facts truly.[22]

Indeed, models' truth*less*ness is universally acknowledged by scientists and social scientists.[23] It might seem that, as models are never "true," they must, at least in some sense, always be false. And, of course, in a narrow sense, they are. Climate change deniers associated with the Heartland Institute and its carbon-dependent corporate funders cynically exploit models' inherent falseness. But rather than confuse our discussion of models by introducing the notions of true and false, we might choose the philosophically simpler alternative of judging models, like we do people, as good or bad, honest or dishonest. Implicit in accepting that good, honest models help us better understand the world is the recognition that there are also bad, dishonest models that, intentionally or unintentionally, distance us from that understanding.[24] The philosopher Harry Frankfurt might even suggest that worse than a model that lies is one that bullshits.[25] Frankfurt observes, relevantly to the present political moment, that a bullshitter is more dangerous than a liar, because a bullshitter, in contrast to a liar, has lost the ability to distinguish the true from the false. Even evil models make claims about revealing the world. Of course, those who recognize the malicious character of such models will learn more from them about the truth of the world (its baseness) than those who are taken in by them. Nevertheless, all models are epistemic exercises offering their makers and observers a fuller engagement with the world. It is, consequently, legitimate to make not only practical judgments about models but also ethical, if not moral, ones. Through the discussion of models' operations, my definition expands:

> A model assumes its interpreter's familiarity with its particular hermeneutic conventions. It can be strong or weak, *good or bad, honest or dishonest. No model can ever licitly make truth claims; nevertheless, good models are epistemic operators that work toward a fuller understanding of the world.*

DISCOURSE

Models' relationship to discourse confirms models' epistemological value and contributes further attributes of the community of models. Models stand in a variety of relationships to texts. Many models depend on texts to fully exercise their agency. An economic pie chart that has no labels or no accompanying account is useless. Without a text, an innocent viewer might think that a beautiful arrangement of glass balls was a delicate sculpture rather than a set of models of the molecules cyclohexane, methane, ethane, and heptane. Other models eschew texts altogether; their meaning can be adequately grasped without the help of a text. The topographic model with which bees negotiate a terrain has nothing whatsoever to do with words.[26] Understanding a supermodel like Naomi Campbell may depend on a familiarity with her cultural context, but it does not rely on an explanatory description. A child requires no information on how to deploy her model truck. At the other extreme are models that are so much like texts that they might be taken to be texts. Algorithms, which are commonly treated as models in the sciences and social sciences, offer examples like this one:[27]

$$\dot{X}_i = \mathrm{f}(X_i) + \sum H(X_j, X_i)$$
$$i = 1, \ldots, 8.$$

An algorithm is not, however, a text but rather an alphanumeric sign. Other models might even be imbedded in texts. Henry David Thoreau's *Walden,* for example, might well function as a model.[28] We understand the assignment when a high school teacher directs her students, "Spend one hour alone closely observing nature. Draft an informal essay or response, modeled on *Walden,* to describe your thoughts and feelings."[29] Similarly, *Walden* is taken as a model, if somewhat vaguely, by B. F. Skinner for the fictional community in his novel *Walden Two.*[30] Texts certainly may function as models for other texts. Texts may also act as models for other things. The high school assignment implies that Thoreau's *Walden* might function as a model for living an ethically reflective lifestyle. It might even be suggested that *Walden* provided a model for Heidegger's hut. Certainly the fictional *Walden Two* is specifically cited as a model for the communities of Twin Oaks in Virginia and Los Horcones in Sonora, Mexico.[31] Of course, to function as a model, the relevant bits of *Walden* must be removed from the text and reconstructed analogically or abstractly as a model or model-diagram. That process of extrapolating a model from the text is its making. In such a case, it might be argued that, though the model is dependent on a text, it is itself not a text but rather a formula, a diagram, an idea.

It is understandable that some texts may be identified as models; nevertheless, models should not be mistaken for texts. Perhaps the most robust example of the identification of a model as a speech form closely affiliated with texts is offered by Max Black, a distinguished philosopher, in his widely cited article "Models and Archetypes," published in 1962.[32] This piece has become the touchstone for many thoughtful theoretical considerations of models, both scientific and humanistic. In it Black defines four different model types (scale, analog, mathematical, and theoretical), but he is really only interested in the last one. Black deftly identifies the theoretical model as a language-thing. "Theoretical models (whether treated as real or fictitious) are not literally constructed: the heart of the method consists in talking in a certain way."[33] Although Black acknowledges that theoretical models are objects, they alone of all model-types are freed from "the controls enforced by the attempt at actual construction."[34] Scale, analog, and mathematical models are constrained by their physicality; theoretical models in contrast, liberated from materiality into language, more effectively promote original thinking. Theoretical models, Black argues, are metaphors.

Models—theoretical or otherwise—are not, however, metaphors. A comparison between "model" and "scaffold" suggests that models are more resistant to Black's textualization than many other things, even spatial ones. A scaffold is a conspicuous, fixed assemblage that enables the construction of a building or the execution of a criminal. To do its work adequately, a scaffold is necessarily stable. But it is also temporary: when its work is finished, the scaffold is dismantled. Its existence may be essential, but it is assumed to be brief. The scaffold is inevitably marginal and readily forgotten. Memories commonly cling to the structure or body once sustained by the scaffold, but only rarely to the scaffold itself. It is an entity without much of a life of its own.

Scaffold and model share significant properties. Both come in a variety of forms, prominent among which are spatial, quasi-architectural structures.[35] Both are open to tinkering and manipulation. Both play a privileged role in the ordering of human physical and mental topographies. But obviously, there are important distinctions between scaffold and model. They perform very differently. Each names a different, although at times overlapping, way of working. The scaffold is always planned, programmatic, and practical; it is applied strategically. The model tends to be correlative, creative, even aleatory; it is deployed tactically. The scaffold can be assumed to do what it is supposed to do; the model tends to be fractious—often frustrating its handler's expectations. Both scaffolds and models can be either conceptual or material, but only models can be simultaneously conceptual *and* material. This distinction may be more fully appreciated in the significantly different ways in which scaffold and model relate to metaphor.

Scaffold is more powerful and less evanescent as a metaphor than it is as a thing. Scaffolds, as metaphors, are ubiquitous, whether describing conference sessions, medical procedures, indexing systems, or discourse analysis and hermeneutics. For example, scaffold "as a metaphor . . . currently seems to be omnipotent in conversations about general education."[36] Certainly architectural terms are familiarly deployed metaphorically. From the Song of Solomon's "a garden enclosed is my sister, my spouse" to Ibsen's *Pillars of Society,* they are everywhere. But apart from "scaffold," not many architectural objects live more fully as metaphors than as objects-in-the-world. Model, in contrast to scaffold, is dysfunctional as a metaphor. "Her work was her scaffold" suggests that labor kept the subject's life upright. "Her work was her model" might signify that labor provided an order that the subject applied to the rest of her life. But the sentence might just as well mean that the subject devoted her life to the construction of a doll's house, like Carrie Stettheimer.[37] Model doesn't work well as a metaphor, because, in contrast to scaffold, model stubbornly bears with it its literalness. No one would claim that a scaffold *is* literally a metaphor, but Black makes that claim about the model. The metaphoricity of scaffold suggests why metaphor's application to model should be more strictly limited.

In the sentence "all the world is a stage," the metaphor is neither "the world" nor "the stage" but rather their synergy: the one makes the reader rethink the other. It is, as Black compellingly argued, the interactivity of the object words that makes the metaphor. Black found a capacity to evoke new ways of thinking about the world in both the theoretical model and the interactive metaphor. He therefore equated them. But a model, in contrast to a metaphor, is not an operation; it is an operator, an agent. It is an entity, not a literary construction. Its actuation of new conceptions of the world is not limited to a particular syntactical setting. It can be manipulated and recontextualized. It is, as Morgan argues, autonomous. That autonomy enables a model to act as a powerful mediator between theory and data (like an economic model) or between the senses and the world (like a map). That autonomy also makes the model resistant to its identification as an ineffable "extension of meaning" that happens between symbolic signifiers in a figure of speech. While it might be metaphorically suggested that a model is a metaphor, it can no more be claimed that a model is *really* a metaphor than it can be claimed that a stage or a scaffold is *really* a metaphor. By identifying models (most particularly theoretical models) as autonomous agents, we frustrate Black's understanding of models as metaphors.

Insofar as a model reforms its object or theory in the process of its construction (conceptual or material), it might act like a metaphor. But the metaphor remains elusive; it makes no testable claims. The metaphor and the model have different epistemic functions. Even when a metaphor is vaguely about "truth," it floats free of contention. Who could argue with the observation that the "world is a stage" or "silence is golden"? A metaphor, like God, is not subject to falsification. But a model, whether material or theoretical, humanistic or scientific, is, as Popper has persuasively argued, the product and object of critique. From Joan Smalls and Barbie to economic pie charts and algorithmic representations of a cantering horse, models are contested and contextually specific.

Why does Black, a very smart philosopher, frame an argument that is unconvincing even to a philosophically naive architectural historian like myself? The answer: he was marshaling his analytic rhetoric to make a more fundamental claim. The theoretical model was Black's Trojan Horse. He used the model, on which science (and the social sciences) were increasingly dependent, to demonstrate that in the sciences, no less than the humanities, imagination, stimulated by creative, often serendipitous juxtapositions of seemingly very different things, is the progenitor of new knowledge. Black used a quintessentially literary device to make scientists (and social scientists) less certain that they are so different from poets and so much more objective than architectural historians. "When the understanding of scientific models and archetypes comes to be regarded as a reputable part of scientific culture, the gap between the sciences and the humanities will have been partly filled. . . . For science, like the humanities, like literature, is an affair of the imagination."[38] My contention that models are not metaphors is not meant as a critique of Black's conclusion that the sciences and humanities are closer than scientists and humanists tend to think. Rather, it supports that claim, though in a different way: the models of humanists, social scientists, and scientists are all members of the same community. Black's claim that models are metaphors has been put to unwonted use by climate change deniers, who use it, often without citing their source, as a means of discrediting findings based on scientific models.[39] My support for Black's appeal that the humanities and the sciences are closer than either realizes might further fuel the idiocy of global warming naysayers.

Not all models have textual relations, but every model engages in discourse. For those of us in the humanities, discourse includes, in addition to linguistic forms, nonverbal things and practices—uniforms, cubicles, airport security, table settings—that articulate and enable the exercise of power relationships. Michel Foucault authorized these accretions by brilliantly describing the participation of vision and the body in the development of the new disciplines of the emergent modern state.[40] Foucault's work has been complemented by that of Bruno Latour, whose actor–network theory (ANT) not only argues for the agency of objects but also persuasively describes the web of human and nonhuman interactions that constitutes the world.[41]

Models may be counted among the most powerful of ANT actors. In the sciences and social sciences, models have become a dominant means of articulating problems, representing theories, and offering descriptions of the world. The scientific model's complex interactions with its makers, observers, and referents in the world are negotiated through a wide variety of recognized media, from robotic data collection to scholarly articles. Scientific discourse is committedly conventionalized. Its quantitative values and qualitative expressions are ordered by broadly accepted symbolic forms and institutional hierarchies. Climate change models are possible not only because of the proliferation of data collection stations in the world but also because of the standardization of the data that are collected. Whose climate change models are taken most seriously as predictors depends on the reputation of the scientists who produce them and the prestige of the institutions that support their research.

In the arts, the humanities, and daily life, the model is not only less privileged but its modes of mediation are less formal. Its discourses are different. Nevertheless, the nonscientific model relates with ease and intimacy to its human producers and consumers through a less ordered but equally complex web of texts, narratives, images, and other objects. The model's capacity to produce stories as well as to illustrate them depends, at least in part, on its singularity. Models are, after all, distinct not only from their referents but also from one another. Each model has its own history. As independent objects in the world, models often outlive those who construct them: models that survive over time continue to function, but in distinct ways in distinct contexts. Models are different than fungible commodities.

Every tomato has a history, but it, like all other fungible commodities, loses that history, along with all its other particularities, when it is processed into ketchup by Heinz. In contrast, models, as autonomous agents, have lives that may with some effort be reconstructed. Models, in other words, have histories.

Any agent that has a knowable history is bound also to have a recoverable politics. The politics of some models is incontrovertible. Models used to predict global climate change are at the center of the political struggle between conservative fossil fuel corporate interests and the rest of us. The politics of many other model-types, from supermodels and weather maps to toy trains and DNA models, are less obvious. But I believe that all models not only have a discursive context and a history but also have a politics, whether that politics is a macropolitics or a micropolitics. The chapters in *Modelwork,* dealing as they do with a variety of particular models, document such an assertion. They demonstrate how model-thinking, along with world-thinking, shifted dramatically over time: they work as evidence of the final claim made in my definition. That definition, for the moment, reads as follows:

> A model is an autonomous agent that has a referent (material, ideal, conceptual, or imaginary) to which it semiotically averts (iconically, indexically, or symbolically) but from which it differs in significant ways (in complexity, scale, material, function, ontology, and the like). A model assumes its interpreter's familiarity with its particular hermeneutic conventions. It can be strong or weak, good or bad, honest or dishonest. No model can ever licitly make truth claims; nevertheless, good models are epistemic operators that work toward a fuller understanding of the world. *All models are entangled in discourse; they have histories, and they act politically.*

NOTES

1 Nelson Goodman, *Languages of Art: An Approach to a Theory of Symbols* (Indianapolis, Ind.: Hackett, 1985), 171–72.

2 Other short definitions of *model* have been offered, but not argued. See, e.g., Luis O. Arata, "A Unified View of Models," *Leonardo* 44, no. 3 (2011): 282–83. His suggestion that a model is "an interface that enables the performance of tasks according to preferences" denies a model its own life.

3 Peirce defined these terms in multiple texts, one of which is Charles Sanders Peirce, *Collected Papers of Charles Sanders Peirce* (Cambridge, Mass.: Belknap Press, 1960), 2.156–73.

4 For a stimulating mathematical definition of a model, see Alain Badiou, *The Concept of the Model: An Introduction to the Materialist Epistemology of Mathematics* (Melbourne: re.press, 2007).

5 Albert Atkin, "Peirce on the Index and Indexical Reference," *Transactions of the Charles S. Peirce Society* 41, no. 1 (2005): 161–88.

6 For the uses of limiting art historical discussion to Peirce's narrower definition of index, see Alexander Robins, "Peirce and Photography: Art, Semiotics, and Science," *Journal of Speculative Philosophy* 28, no. 1 (2014): 1–16.

7 Martin Heidegger, "The Thing," in *Poetry, Language, Thought,* ed. Albert Hofstadter, 163–87 (New York: Harper and Row, 1971); Bill Brown, "Thing Theory," *Critical Inquiry* 28, no. 1 (2001): 1–22.

8 Annabel Wharton, *Architectural Agents: The Delusional, Abusive, Addictive Lives of Buildings* (Minneapolis: University of Minnesota Press, 2015). I have also used it elsewhere, notably in Annabel Wharton, "Scaffold, Model, Metaphor," *ARPA* 4 (May 2016), http://www.arpajournal.net/scaffold-model-metaphor/.

9 William C. Anderson, *A Dictionary of Law, Consisting of Judicial Definitions and Explanations of Words, Phrases, and Maxims, and an Exposition of the Principles of Law* (1889; repr., Union, N.J.: Lawbook Exchange, 1996). For a more modern, more detailed, but less cogent legal definition of *agent,* see Jonathan Law and Elizabeth A. Martin, *A Dictionary of Law* (Oxford: Oxford University Press, 2018).

10 In Mary S. Morgan and Margaret Morrison, eds., *Models as Mediators: Perspectives on Natural and Social Science* (Cambridge: Cambridge University Press, 1999), chapter 3. For a brilliant analysis of economic models and their histories, also see Mary Morgan, *The World in the Model: How Economists Work and Think* (Cambridge: Cambridge University Press, 2012).

11 Rick Ferri, "The Curse of the Yale Model," *Forbes,* April 16, 2012. I am indebted to Jessica Newman, who wrote a fine paper on this subject for my Models Seminar at Duke in spring 2015.

12 Clifford Geertz, "Religion as a Cultural System," in *The Interpretation of Cultures: Selected Essays* (1973; repr., London: Fontana Press, 1993), 93.

13 Dictionary.com, s.v. "ethics," http://www.dictionary.com/browse/ethics.

14 Morgan, *World in the Model,* 38.

15 Morgan and Morrison, *Models as Mediators,* 11–12.

16 Marcial Echenique, "Models: A Discussion," *Journal of Architectural Research* 1, no. 1 (1970): 27.

17 Johan Huizinga, *Homo Ludens: A Study of the Play Element in Culture* (New York: Roy, 1950), 206.

18 Some of Huizinga's stipulations for play are less applicable: its amorality and impracticality.

19 Huizinga, 10.

20 Karl Popper, *The Myth of the Framework: In Defence of Science and Rationality* (London: Routledge, 1994), 174.

21 Popper, 173–74.

22 Popper, 172–73.

23 Kevin A. Clarke and David M. Primo, *A Model Discipline: Political Science and the Logic of Representations* (New York: Oxford University Press, 2012); Gabriele Contessa, "Scientific Models and Representation," in *The Continuum Companion to the Philosophy of Science,* ed. Steven French and Juha Saatsi, 120–35 (New York: Continuum, 2011).

24 See Henry Ergas, "For a Charter of Modelling Honesty," *Agenda* 19, no. 2 (2012): 5–8.

25 Harry G. Frankfurt, *On Bullshit* (Princeton, N.J.: Princeton University Press, 2005).

26 James F. Cheeseman, Craig D. Millar, Uwe Greggers, Konstantin Lehmann, Matthew D. M. Pawley, Charles R. Gallistel, Guy R. Warman, and Randolf Menzel, "Way-Finding in Displaced Clock-Shifted Bees Proves Bees Use a Cognitive Map," *Proceedings of the National Academy of Sciences of the United States of America* 111, no. 24 (2014): 8949–54.

27 As evident in this title: William Sherman and Anna Grosberg, "An Adapted Particle Swarm Optimization Algorithm as a Model for Exploring Premyofibril Formation," *AIP Advances* 10, no. 4 (2020), https://doi.org/10.1063/1.5145010.

28 Henry David Thoreau, *Walden* (Princeton, N.J.: Princeton University Press, 1971).

29 Tonya Dehorn, "Los Gatos High School, Assignments for English 11: September 21–25," 2015.

30 B. F. Skinner, *Walden Two* (Indianapolis, Ind.: Hackett, 1948).

31 Hilke Kuhlmann, *Living Walden Two: B. F. Skinner's Behaviorist Utopia and Experimental Communities* (Champaign: University of Illinois Press, 2005).

32 Max Black, "Models and Archetypes," in *Models and Metaphors: Studies in Language and Philosophy,* 219–44 (Ithaca, N.Y.: Cornell University Press, 1962).

33 Black, 229.

34 Black, 229.

35 Or now, with biological nanotechnologies, the building of body parts. Christopher L. Dearth, Timothy J. Keane, Jeffrey R. Scott, Kerry A. Daly, and Stephen F. Badylak, "A Rodent Model to Evaluate the Tissue Response to a Biological Scaffold When Adjacent to a Synthetic Material," *Tissue Engineering Part A* 21, no. 19–20 (2015): 2526–35.

36 Nancy Boblett, "Scaffolding: Defining the Metaphor," *Columbia University Working Papers in TESOL and Applied Linguistics* 12, no. 2 (2012): 1–16.

37 Annabel Wharton, "Doll's House/Dollhouse: Models and Agency," *Journal of American Studies* 53, no. 1 (2017), https://doi.org/10.1017/S0021875817000895.

38 Black, "Models and Archetypes," 243.

39 Craig D. Idso, Robert M. Carter, and S. Fred Singer, *Why Scientists Disagree about Global Warming: The NIPCC Report on Scientific Consensus* (Arlington Heights, Ill.: Heartland Institute, 2016); Jerry Ravetz, "Models as Metaphors," in *Public Participation in Sustainability Science,* ed. Bernd Kasemir, Jill Jäger, Carlo C. Jaeger, and Matthew T. Gardner, 62–77 (Cambridge: Cambridge University Press, 2003).

40 For a fuller historiography of this intellectual development, begin with Stuart Hall, "The Work of Representation," in *Representation: Cultural Representations and Signifying Practices,* ed. Stuart Hall, 13–74 (Thousand Oaks, Calif.: Sage, 1997).

41 Bruno Latour, *Reassembling the Social: An Introduction to Actor-Network-Theory,* Clarendon Lectures in Management Studies (Oxford: Oxford University Press, 2005).

2

Material Models of Immaterial Things

PETER GALISON

There is a dog-bites-person story about models that we all know very well. Faced with a complex physical system, we invoke a stripped-down system of abstractions to explain it. We recognize that Earth is not spherically symmetric. Yet assuming that it is composed of a series of concentric shells offers a tremendous simplification, one good enough for many purposes. We know that crystals are not geometric but reason successfully with cubic structures; we recognize that fluids in many circumstances are somewhat compressible but build descriptions of them with incompressible fluids to account for many phenomena.

This form of representational abstraction is at the center of so much of the continuing revolution, set in motion by Descartes, Galileo, and Newton, that it has become second nature to us—invoking abstractions of these ideal types to get at our messy, worldly reality is in many cases just what we mean by explanation.[1] Models like these—idealizations—seem to be at the heart of this enterprise of explanation.[2]

Once in a while, far less often than such "normal" cases, we encounter a form of modeling that seems, on the face of it, to be utterly backward: the modeling of an abstract, immaterial entity by a material system. Why would one use a physical entity, a real-world concatenation of gears and springs, of flowing water in rubber tubes, of straps, springs, and pulleys, to model something that itself was supposed to be captured by differential equations? At first sight, it might seem that a material model is "merely" an aid to the feeling of having grasped a purely theoretical account. Indeed, there is a philosophical-scientific tradition that takes understanding itself to be nothing more than a pedagogical, popularizing, or psychological crutch.[3]

To see the value of material models, we need a shift of priorities: from the near-ubiquitous hunt for explanation to the less-attended-to epistemic virtue of understanding. Years ago, Michael Friedman took a shot at adjudicating the relationship of understanding to explanation, depicting understanding as important—not just a decorative complement. But he kept understanding tied to explanation. Indeed, he saw in the literature three ways of making the link from explanation to understanding. One way of specifying understanding is to say that understanding is present when a phenomenon can be deduced from another more securely grasped range of phenomena. A second passage from explanation to understanding takes place when we can show that some novel phenomenon can be tied back to a familiar one (even if not deductively), for example, we explain thermodynamic laws in terms of the collision of miniature billiard balls. Yet a third school says that understanding is present when it relates new phenomena to some historically specific (changing) set of prevalent conceptions. Friedman's own view is that all three notions of understanding fail because all involve a *local* explanation of something new in terms of something accepted. By contrast, Friedman's own position was that we understand something when we *unify* it vertically into the comprehensive, global set of scientific conceptions such that "our over-all understanding of the world is increased; our total picture of nature is simplified via a reduction in the number of independent phenomena."[4] The view that explanations must be present to have understanding—or even that understanding is to be identified with explanation—remains widespread.

A more promising approach—as I see it—is taken by Henk de Regt, who values understanding as an epistemic goal in its own right. He identifies understanding (applied to theories) as closely allied with *intelligibility*: the features of a theory that are taken in a historical period to facilitate the theory's use.[5] This is in line with Jordi Cat's approach to James Clerk Maxwell's use of metaphor and illustration: they help establish intelligibility, and with that intelligibility comes a precondition for explanation. "Understanding then establishes the cognitive basis for explanatory value."[6]

Like de Regt, Cat, and a growing number of others, I want to focus on understanding—though even less than they do on explanation and more on physical modeling and its links with other approaches to knowledge. Understanding, I will argue here, lies in this *coordination without homogenization*.[7] As one can imagine, the etymology of *understand* is much discussed and disputed, but one strand, the one I find most intriguing, carries with it the vestige of literally "standing among." This is preserved not only in the German *verstehen* (stand among), both and in other languages, too, deriving from Proto-Indo-European **nter,* designating "between" or "among" (not just literally "beneath").[8] But etymologically bolstered or not, it is this sense of the concept that seems most central to me: to understand is to stand among—and eventually to advance among—forms of knowing, drawing together the symbolic, the visual, and the material.

From quantum field theory to the British economy, from the nineteenth-century all-pervasive electromagnetic ether to contemporary accounts of black holes, I would like to explore some of these surprising moments of modeling the abstract by the concrete.[9] The goal is to grasp this back-and-forth between material models and immaterial things, the lateral correspondence that makes up at least this one, often peculiar part of what we mean by understanding.[10]

SYMBOL, SPACE, AND BODY

In his "Dynamical Theory of the Electromagnetic Field" (1864), James Clerk Maxwell turned from the examination of his twenty equations (of electrical and magnetic force, resistance, continuity, and more) to phenomena solidly within the domain of gears and pulleys. In the past, he had spoken of real strains and motions in the ether. Here he wanted to be more abstemious: "I wish merely to direct the mind of the reader to mechanical phenomena which will assist him in *understanding* the electrical ones." All such invocations were, he insisted, "illustrative" and not "explanatory."[11]

In September 1870, in his Liverpool Presidential Address to the British Association, Maxwell reflected on the different kinds of people who reason about the physical world. True, the scientist is willing to act like a "calculating machine" for a time. Such procedural manipulations can make things clearer. But the act of one justified step after another can leave one without a synoptic vision of the whole; there are times when the steps are so numerous that a person "is sure to forget before he has reached the conclusion." That is the moment to put machine-like reasoning aside and turn elsewhere: "to *understand* the subject by means of well-chosen illustrations derived from subjects with which he is more familiar." Such "scientific illustrations" are, the Victorian polymath contended, invaluable. They "enable the mind to grasp some conception or law in one branch of science, by placing before it a conception or a law in a different branch of science, and directing the mind to lay hold of that mathematical form which is common to the corresponding ideas in the two sciences, leaving out of account for the present the difference between the physical nature of the real phenomena." In this way, one could attain knowledge "more profound than could be obtained by studying each system separately."[12]

True, Maxwell noted, there were those who, faced with a mathematical relation or law, no matter how complex, could grasp, purely from the symbolic, abstract representation alone, the full meaning carried by formal symbols. "Such men," Maxwell noted, "sometimes treat with indifference the further statement that quantities actually exist in nature which fulfil this relation. The mental image of the concrete reality seems rather to disturb than to assist their contemplations." Most of humanity, however, could not, absent a particular training, seize these relations, much less "retain in their minds the unembodied symbols of the pure

Equations of Electromotive Force.

$$\left.\begin{aligned} P &= \mu\left(\gamma\frac{dy}{dt} - \beta\frac{dz}{dt}\right) - \frac{dF}{dt} - \frac{d\psi}{dx} \\ Q &= \mu\left(\alpha\frac{dz}{dt} - \gamma\frac{dx}{dt}\right) - \frac{dG}{dt} - \frac{d\psi}{dy} \\ R &= \mu\left(\beta\frac{dx}{dt} - \alpha\frac{dy}{dt}\right) - \frac{dH}{dt} - \frac{d\psi}{dz} \end{aligned}\right\} \ldots\ldots\ldots\ldots\ldots\ldots\text{(D)}.$$

FIGURE 2.1. Maxwell's equations of electromotive force, from "A Dynamical Theory of the Electromagnetic Field, 1864," in *The Scientific Papers of James Clerk Maxwell,* ed. W. D. Niven (1890; repr., New York: Dover, 1965), 1:559.

mathematician." Indeed, if science was to "become popular, and yet remain scientific, it must be by a profound study and a copious application of those principles of the mathematical classification of quantities which, as we have seen, lie at the root of every truly scientific illustration." For Maxwell, to render physics "popular" did not imply merely bringing these relations to the masses. Instead, making the discipline popular meant altering the science for physicists who were not, could not, be satisfied with differential equations alone.

Of course, Maxwell allowed that "there are . . . some minds which can go on contemplating with satisfaction pure quantities presented to the eye by symbols, and to the mind in a form which none but mathematicians can conceive."[13] Maxwell believed that the pure algebraic-mathematical mind found the physical instantiation of an equation to interfere with their "contemplation." No better example comes to mind of these complex abstractions than the equations that now bear Maxwell's name—among them his expression for the electromotive force for a moving conductor (Figure 2.1).

But there were other ways to grasp the physical situation that did not depend exclusively on differential equations. This second kind of mathematical mind, also thrived in abstraction, but in a way that was not at all algebraic. "Others," he continued, "feel enjoyment in following geometrical forms, which they draw on paper, or build up in the empty space before them."[14] Maxwell drew such forms to illustrate how the lines of force could be derived from the contours of constant values of the potentials (Figure 2.2).

Third, and in Maxwell's categorization, finally, were those among the scientists who had a different conception altogether, one that Maxwell described rhapsodically:

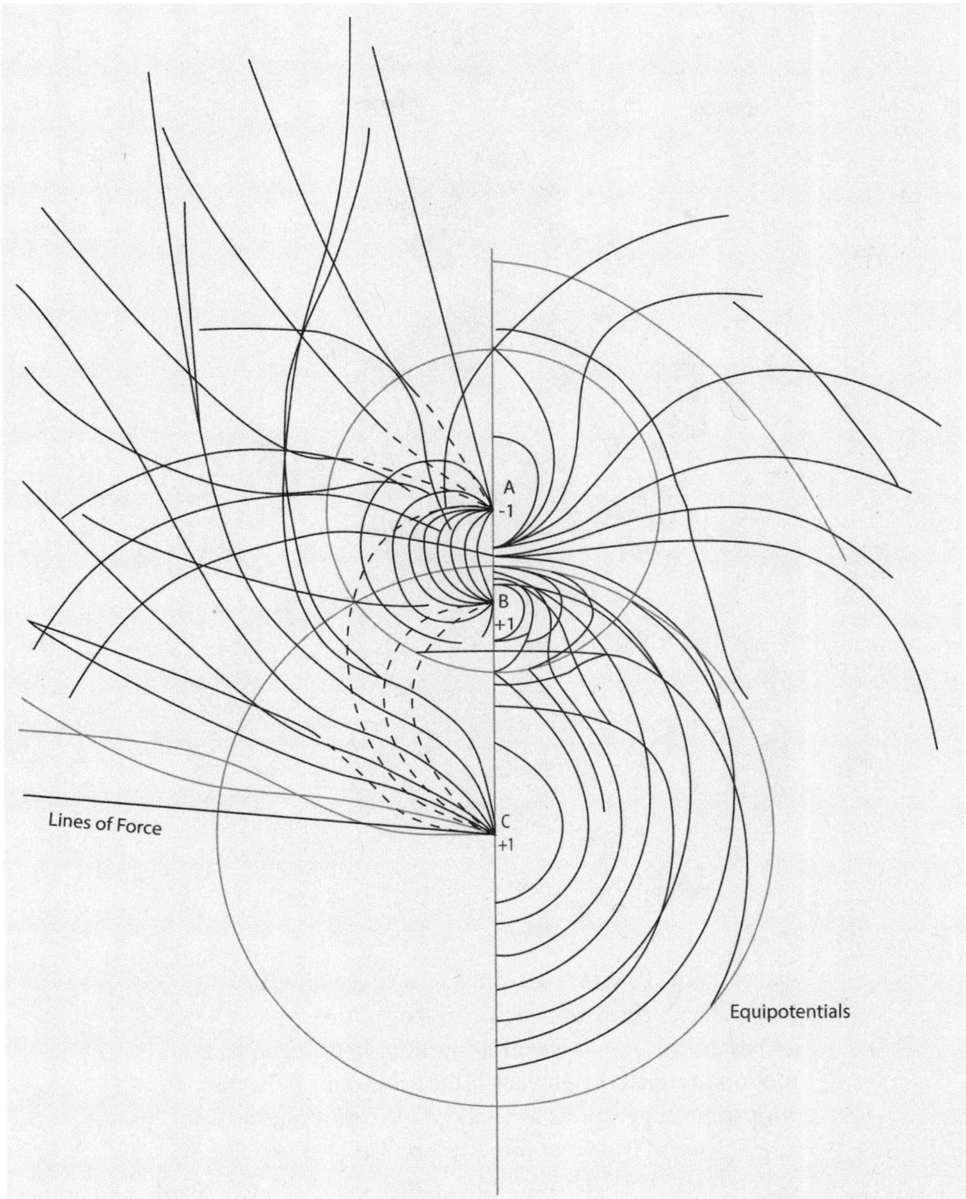

FIGURE 2.2. Maxwell's geometry of force. Two spheres with opposite charges: A, impermeable, B, a perfect conductor. Forces are orthogonal to the contours of equal potential.

"Others . . . are not content unless they can project their whole physical energies into the scene. . . . They learn at what a rate the planets rush through space, and they experience a delightful feeling of exhilaration. They calculate the forces with which the heavenly bodies pull at one another, and they feel their own muscles straining with the effort." Maxwell wrote of those who wanted this feeling of

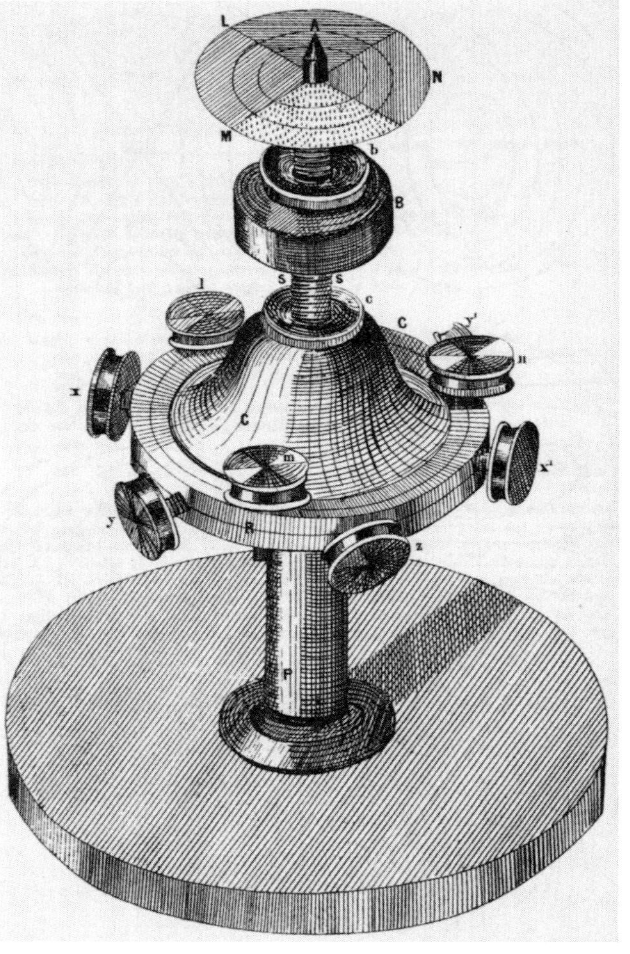

FIGURE 2.3. Poinsot's astronomical precession in a toy. Maxwell exhibited an instrument/toy of this type in "On a Dynamical Top, for exhibiting the phenomena of the motion of a system of invariable form about a fixed point, with some suggestions as to the Earth's motion, 1857," in *The Scientific Papers of James Clerk Maxwell,* ed. W. D. Niven (New York: Dover, 1965), 1:262.

empathetic bodily response: "To such men momentum, energy, mass are not mere abstract expressions. . . . They are words of power, which stir their souls like the memories of childhood."[15]

One thinks here of Maxwell's own work echoing in many of his studies. In April 1857, he turned to the dynamical top, returning again to the image of a childhood contemplation of matter in motion. "The mathematicians of the last age, searching through nature for problems worthy of their analysis, found in this toy of their youth, ample occupation for their highest mathematical powers." Indeed, as he insisted, "no illustration of astronomical precession can be devised more perfect than that presented by a properly balanced top." Here he credited French physicist and mathematician Louis Poinsot (who had shown how forces on a rigid body could be captured by a single force and couple) for having brought to the subject the "power of a more searching analysis" than anything mere calculus could provide (Figure 2.3). "Ideas take the place of symbols, and intelligible propositions supersede equations."[16]

This toy-instrument—a form of illustration in Maxwell's sense—was key to understanding and a source of pleasure. Sigmund Freud once said that the original, true, and greatest happiness was the fulfillment of a childhood desire, the reengagement of that desire fulfilled in focused, serious, committed play. Of course, says Freud, the form this play takes is ever-shifting with circumstances: the pleasure of concentrated play shifts over time, but it always has a "date-mark" on it.[17] So it seems to be for Maxwell, as he invoked the pleasurable "memories of childhood" associated with bodily expression drawn up into concepts. Maxwell's *Traum* not *Trauma* found its site in the attachment of abstractions—wound up in momentum, energy, mass—to real kinesthetic, muscular, *physicalized* action, often coming to ground, so to speak, in the manipulation of toys–become–physicalized illustrations.

In April 1879, when Maxwell examined the electrical potential, he wanted something one could feel—the way one could (so he contends) feel pressure by plunging into the depths of the sea in a diving bell or feel temperature in a Turkish bath. But electrical force seemed mired in abstraction, referring to the relation of quantities measured at two different points; he worked insistently to afford a way to engage kinesthetically with electrical force. "It is *modelled* on the lines of the familiar definition of mechanical force, and those who find that they understand mechanical force better by 'feeling what it is like' can easily apply the same method to the study of electromotive force."[18]

It was just this feeling "what it is like" that Maxwell was after when, in 1858, he built a physical model designed to illustrate why the rings of Saturn had the form and stability they did (Figure 2.4). Again, physical modeling by *felt* mechanical force lay central to the struggle to understand. "Illustrations," for Maxwell, therefore had a technical specificity: they were not just the kind of one-off metaphors that swarm popular science. Instead, as Maxwell put it in 1870, "the correctness of such an illustration depends on whether the two systems of ideas which are compared together are really analogous in form, or whether, in other words, the corresponding physical quantities really belong to the same mathematical class."[19] For Maxwell, this congruence, under a common sheltering mathematical structure, was what led to a deeper understanding than any particular physical realization could provide.

This class relation, this relation of different realizations, is what I have in mind with the idea of a *lateral coordination* of approaches to a phenomenon—each differently articulates that which is investigated, for example, the rings of Saturn or the relation of magnetism to electricity. Together, the various modes deepen understanding by multiplying the ways of carrying on (rather than any form of vertical explanation by way of unification or reduction).

Or put another way, having different realizations of knowledge (algebraic, geometric, physical) with a mathematical commonality deepens understanding. The "more profound" knowledge might include but is not necessarily a matter of prediction or explanation. It is something else because it can be *generative,* productive of new forms of scientific work—not just the quantitative anticipation of the behavior of a physical system, nor even the particular explanation of why something occurred. A computer simulation could well produce a very precise quantitative prediction without, for that, creating new domains of application. Indeed, it is this eruption of the new out of the joining of two or more systems that signals the real power of understanding or, for our purposes here, modeling. Sometimes these various modes of thought combine in one person. Maxwell, after all, reasoned by turn geometrically, formally, physically. But, as he suggests, the power of this multimodel understanding can occur in the community, the assembly of scientists who know differently, focusing, like a lens collecting light, their various styles of knowing.

GENERATIVITY

Maxwell himself rejected the idea that the "mind of man" is usefully likened to a computation. The mind (for Maxwell) is not similar to Fourier's heated body, "settling down into an ultimate state of quiet uniformity" that with the right laws one could have anticipated. No, Maxwell insisted, the scientific mind is more like a tree adapting to circumstances, "shooting out branches" toward the sky, plunging roots down into the soil. It is only given to us to "breathe . . . the spirit of our own age," to seize "the characteristics" of our "contemporary thought." Just to be prepared for these *un*-anticipated developments, it is crucial that mathematics and physics sort out which ideas from one domain of physics can be deployed in another.[20]

Maxwell called the "figure of speech" of thought and language that effects this correspondence among various "departments" of science "scientific metaphor" in just those cases where each term of the metaphorical structure keeps the formal relations across departments. When that formal correspondence works, the method is "truly scientific," by which Maxwell means "capable of generating science in its turn."[21]

FIGURE 2.4. Maxwell's 1858 model of the movements of satellites that make up the rings of Saturn, from "Letter to William Thomson, 30 January 1858." Courtesy of and copyright the Cavendish Laboratory, University of Cambridge.

Bruce Hunt nicely captured Maxwell's ambition for models—as a means of broadening the range of acceptable paths forward: through instruction but also original research. When Maxwell vaunted illustrative models as "convenient for teaching science in a pleasant and easy manner," one might take this to mean that they were *only* of pedagogical interest. That is, one might (Maxwell does not) think they were effective in *transmitting* knowledge but not think of them as having epistemic virtue in and of themselves: valuable for gaining, securing, and further advancing knowledge. Such a reduction to teaching or popularizing is clearly *not* what he meant. Indeed, for Maxwell, the illustrative model offered a path for moving mathematical structures from one site to another—around a common physical interpretation—without sacrificing force of argument: "If science is ever to become popular, and yet to remain scientific," there had to be illustrations of this type. It must be by the use of physical illustrations, for these provided the only means by which many areas of physics could be presented in a rigorous yet understandable way to those for whom a bare set of equations would be wholly opaque.[22]

For the Maxwellian Oliver Lodge, like Maxwell, it was the overarching mathematical regularities that bound the flow of material liquids to the flow of immaterial electricity. Electricity, he asserted, "behaves like a perfect and all-permeating liquid. Understand I by no means assert that electricity is such a fluid or liquid; I only assert the undoubted fact that it behaves like one, i.e. it obeys the same laws." Of course (says Lodge), we must be on the lookout for discrepancies, but absent some variation, we ought to pursue the joint path. Should we resist the use of analogies, "there are only two courses open to us: either we must become first-rate mathematicians, able to live wholly among symbols and dispensing with pictorial images and such adventitious aid," or—second option—we give up trying to grasp "the present state of electrical knowledge."[23]

This allusion to becoming a "first-rate mathematician" was no right-handed criticism, so to speak. Although Lodge himself could follow mathematical reasoning, he never contributed to the more formal side of physics in any way. He himself was among those who needed the analogies; they were addressed to him and other leading researchers like him, not just to the great mathematically unwashed.[24] Take Lodge's "cogwheel ether," where the moving rack represented current and the gears set in motion stood for the ethereal whirls that constituted the magnetic field (Figure 2.5). Indeed, just because Lodge was willing to give up his understanding (his "grasp") of modern electricity, he turned again and again to the illustrative-analogical: "Think of electrical phenomena as produced by an all-permeating liquid embedded in a jelly," he says, "think of conductors as holes and pipes in this jelly, of an electrical machine as a pump, of charge as excess or defect, of attraction as due to strain, of discharge as bursting, of the discharge of a Leyden jar as a spring."[25]

FIGURE 2.5. Oliver Lodge's cogwheel ether. The toothed bar, pulled through the apparatus of gears, is analogous to a current, and the cogwheels (labeled + and −) stand in for a magnetic field generated by the passage of that current. From Lodge, *Modern Views of Electricity* (London: Macmillan, 1907), 212.

Electricity could be grasped in terms of these mechanical concepts: strain, burst, jelly, gears, and whirls.

Lodge had in mind fully realized, *physical* models—he warned the reader to make sure the glass volume was fully filled, that all interfering bubbles had been duly pumped out (Figure 2.6). This "conspicuous," manipulatable construction advanced understanding. By following the physical, the analogous Lodge explained to his readers, "you will have made a step in the direction of the truth, but I must beg you to understand that it is only a step; that what modifications and additions will have to be made to it before it becomes a complete theory of electricity I am unable fully to tell you. I am convinced they will be many, but I am also convinced that it is unwise to drift along among a host of complicated phenomena without guide

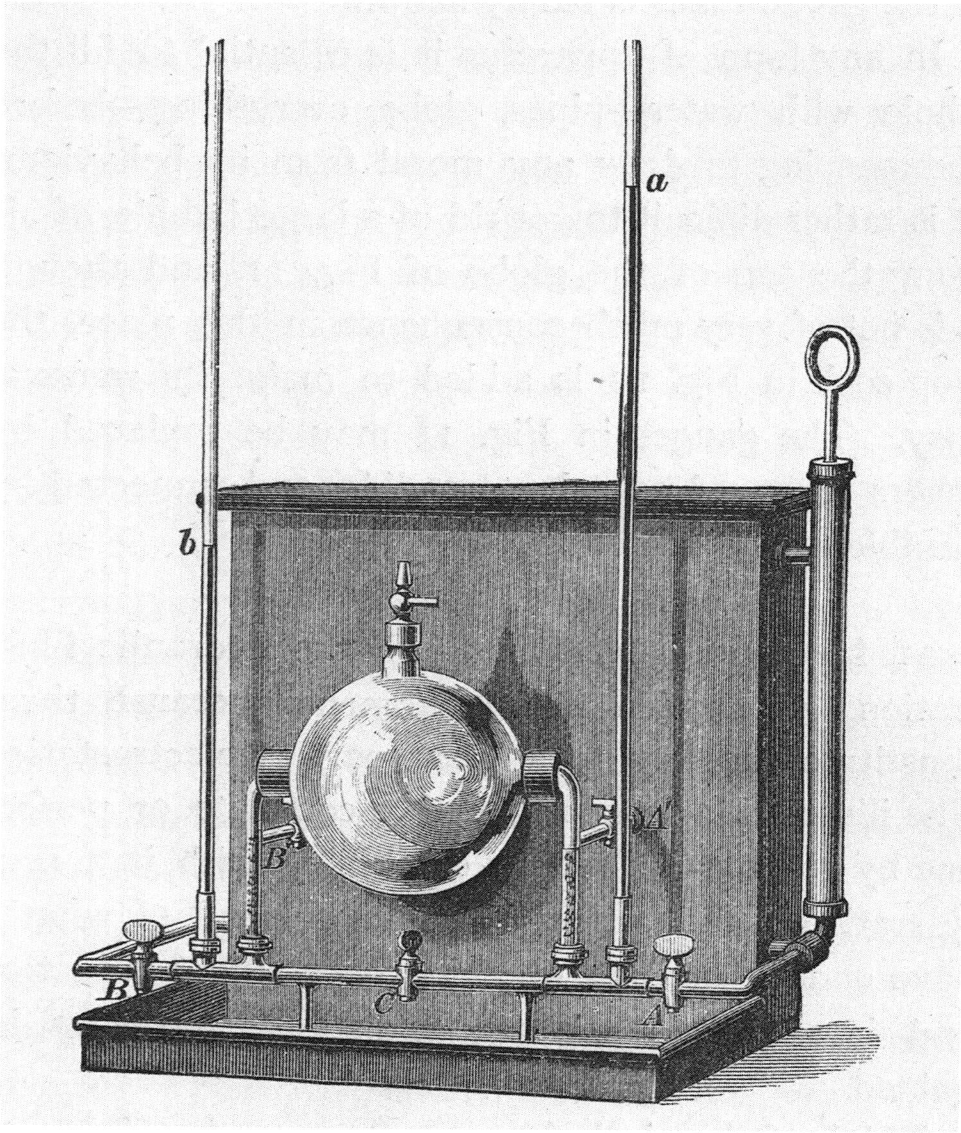

FIGURE 2.6. Oliver Lodge's fluidic-electricity; here a hydraulic model of a Leyden jar: electricity as water and gel. The whole, including the water gauges, is arranged vertically to be "more conspicuous." From Lodge, *Modern Views of Electricity* (London: Macmillan, 1907), 59.

other than that afforded by hard and rigid mathematical equations."[26] Models—physical, moving, material models—offered a path to a generative understanding that complemented the mathematical.

Laboring year after year, decade after decade, to capture the immaterial in the material, Lodge came ever more to see in the ether itself a form of substantiality.

In 1909, he issued an even more ambitious volume, *The Ether of Space.* As its frontispiece, he chose to depict the experimenter working from inside a massive analog device he dubbed "The Ether Machine." By this time, he had come to think of the "insubstantial" ether to be anything but that. "I am able to advocate a view of the Ether which makes it not only uniformly present and all-pervading, but also massive and substantial beyond conception. It is turning out to be by far the most substantial thing—perhaps the only substantial thing—in the material universe. Compared to ether the densest matter, such as lead or gold, is a filmy gossamer structure; like a comet's tail or a milky way, or like a salt in very dilute solution."[27]

Insubstantial—and most massive . . . beyond conception . . . the only substantial thing in the material universe. One thinks of Karl Marx's evocative phrase "all that is solid melts into air." Indeed, for Lodge, all that was material morphed into the immaterial, but conversely, all that had been immaterial had condensed into the most material stuff in all creation. In that back-and-forth, metaphorically, lies the evocative power of the material model. In the material model came a different, but coordinated, way of manifesting the mathematics.

AN ECONOMY OF WATER

Before World War II, New Zealander A. W. H. (Bill) Phillips had been a hands-on engineer trained in the electrical arts and worked at the remote Lake Waikaremoana hydroelectrical plant on the North Island—a bravura mix of electrical and hydraulic engineering.[28] When war broke out, he serviced airplanes, adjusting and modifying guns, among other things. After the war, Phillips headed to England for study at the London School of Economics. Facing lectures on economic theory, he found their abstract character "difficult to understand" and began sketching hydraulic relations. With Walter Newlyn, Phillips began thinking about the machine as a way of instantiating and making visible differential equations through a plumbing-based hydraulic model, in part to facilitate his own learning—an episode well documented (in much greater detail than is possible here) by historian of economics Mary Morgan.[29]

Their ambitious project: to model the British economy with a connected set of water-carrying tubes, stopcocks, and reservoirs that could be controlled and adjusted. To build a device that could do this, Phillips and Newlyn bought war surplus hydraulic pumps originally destined to drive the landing gear, bomb bay doors, and aircraft fuel tanks. One pump had originally been designed for the reserve tank of a Spitfire. For the clear reservoirs, he cut Perspex (also known as Lucite or Plexiglass) sheets from the surplus windows of bombers (Figure 2.7).[30]

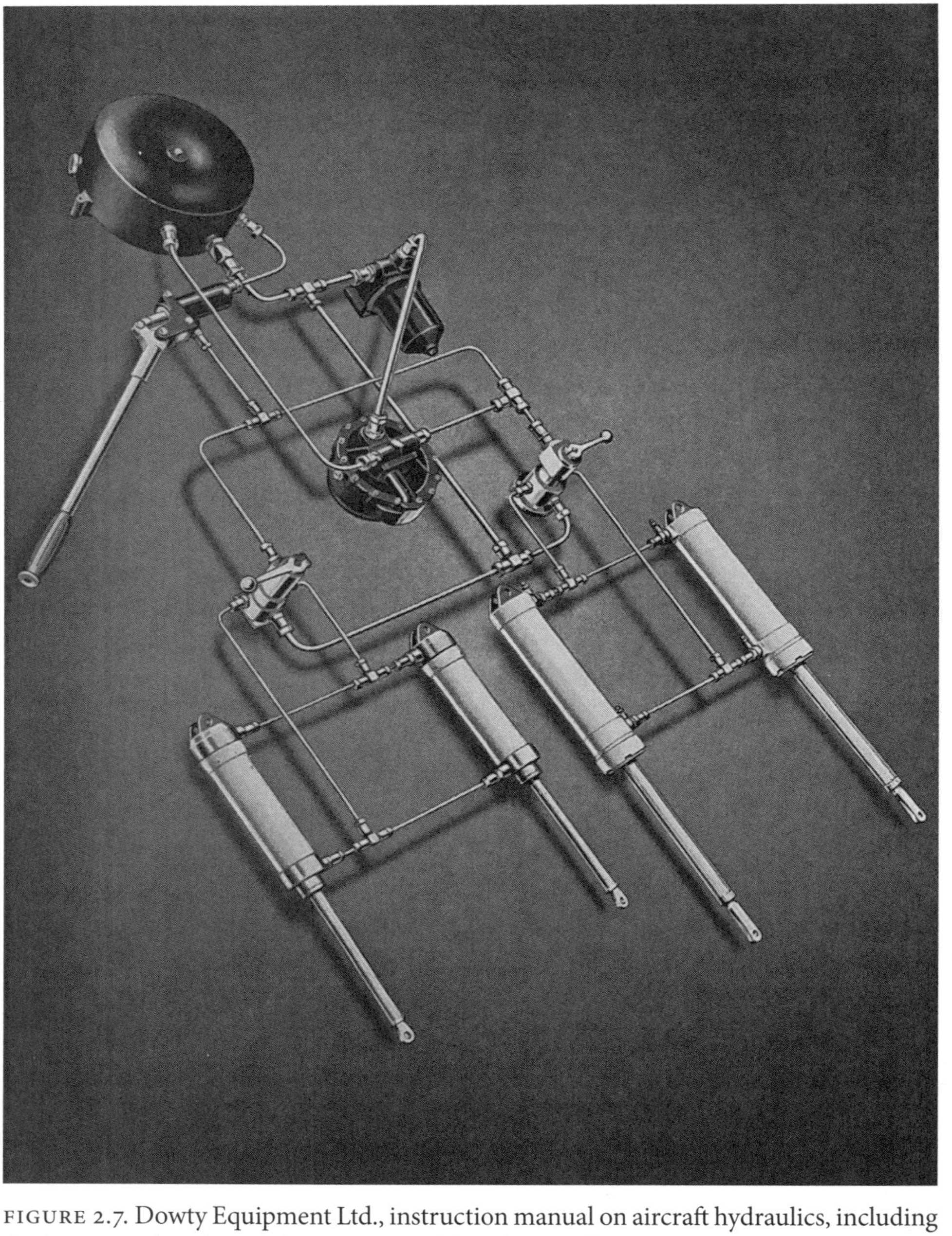

FIGURE 2.7. Dowty Equipment Ltd., instruction manual on aircraft hydraulics, including the Lancaster bomber undercarriage and bomb door, flap control unit, and emergency air shuttle valve. From https://www.dowtyheritage.org.uk/wp-content/uploads/2019/02/Dowty-Manual-1940-2-658x494.jpg.

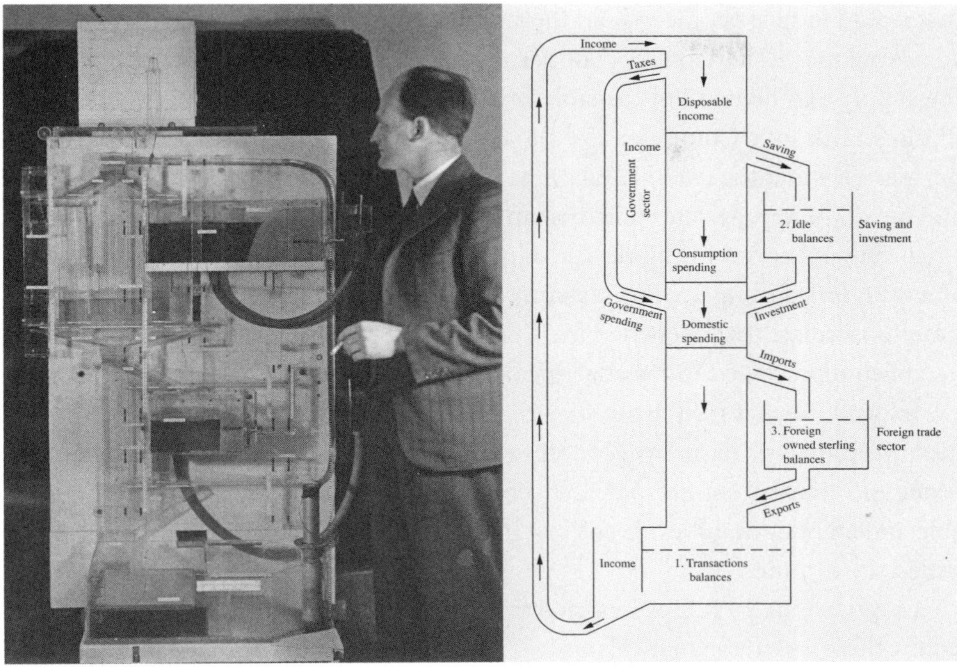

FIGURE 2.8. *Left,* Phillips and his machine—a photograph and a schematic. London School of Economics, circa 1950, https://www.sciencemuseum.org.uk/sites/default/files/styles/embedded_image/public/2018-12/Bill_Phillips_Machine_LSE.jpg. *Right,* a simplified view of the Phillips machine, from Nicholas Barr, "The Phillips Machine," *LSE Quarterly* 2, no. 4 (1988): 321. Used with permission.

(The Lancaster bomber, from which Phillips was scavenging surplus materials, held a very complex hydraulic system, controlling landing gear, bomb bay doors, and more—with every part of the plane regularly subject to wartime stresses and battle damage.)

By November 1949, Phillips was ready to present his machine to the London School of Economics seminar. Seemingly against the odds, "both Phillips and the machine acquitted themselves well. Everyone who mattered was there. . . . Some [came] mainly to laugh. They gazed in wonder at this large 'thing' in the middle of the room" (Figure 2.8).[31] Wonder indeed. Economic debates circulated around differential equations, data, and abstract models. The assembled regarded the pumping, gurgling machine with more than a little doubt. As for the speaker, "Phillips, chain-smoking, paced . . . explaining it in a heavy New Zealand drawl, in the process giving one of the best lectures on Keynes and Robertson . . . anyone . . . had heard. He then switched the machine on. And it worked! He really had created a machine that would simplify the problems and arguments economists had been having for years." According to Keynes, the equilibrium interest rate would take on the value that would induce people to hold the amount of money and number of bonds that were available. Robertson, by contrast, argued that the interest rate would be set by the supply and demand of loanable funds. The MONIAC, also referred to as the Phillips Hydraulic Computer and the Financephalograph, displayed visually and persuasively that the two formulations both held in equilibrium, water flowing in and out of the appropriate containers in steady measure.[32] Water flow embodied the equations and ended the battle. As Mary Morgan put it well, "the Machine enabled a new understanding. . . . The system was genuinely dynamic—the liquid did take time to circulate. [It] deepened their understanding of the economic system that had been represented in the analogical model."[33]

In print, Keynes's much more mathematical formulation and Robertson's more discursive analysis for many economists floated past one another. Put into colored water and transparent Perspex, they could be understood, grasped without being able, independently, to reside entirely in the world of the equations. Flowing water stilled the dispute.

In August 1950, Phillips reported that "there has been an increasing use in economic theory of mathematical models, usually in the form of difference equations, sometimes of differential equations, for investigating the implications of systems of hypotheses. However, those students of economics who, like the present writer, are not expert mathematicians, often find some difficulty in handling these models effectively." Phillips judged that the mechanical models played a crucial role by helping "non-mathematicians by enabling them to see the quantitative changes that occur in an inter-related system of variables following initial changes in one or more of them."[34] A set of differential equations might capture these interrelated quantities, but *seeing* them could be crucial for an economist who was not simultaneously a mathematician—a professional not inside the charmed circle who lived and breathed in a purely symbolic world. In short, his aim to embody abstract relations mirrored those of Maxwell and Lodge, though eight decades later and in an entirely different discipline.

A few years later, Lionel Robbins, a leading member of the economics department at the London School of Economics, recalled the triumph of that moment: how the machine put into motion savings, investment, money supply, and liquidity. Suddenly embodied and controllable, "the subject of extensive debate for the preceding decade, resolved themselves almost automatically."[35] Here was more than a pedagogical supplement. It was, instead, an advance, settling an ongoing abstract-economic debate in tactile-visible form.

Phillips looked at the machine as an alternate way of embodying the reality of a set of equations: the underlying mathematics was realized once in the actual flow of money, and at the same time in the flow of water: "fundamentally, the problem is to design and build a machine the operations of which can be described by a particular system of equations which it may be found useful to set up as the hypotheses of a mathematical model, in other words, a calculating machine for solving differential equations." Other realizations were possible, to be sure. But, Phillips continued, "since . . . the machines are intended for exposition rather than accurate calculation, a second requirement is that the whole of the operations should be clearly visible and comprehensible to an onlooker."[36] With an electrical engineering background, certainly Phillips could have pursued a purely (analog) electrical computer (later in life, he did work with electrical analog computers). As a matter of prediction, it would certainly have been just as good, perhaps better, electrons being more controllable than water.

But instead, Phillips chose water flowing in Perspex. Like Maxwell and Lodge, Phillips turned to embodied abstraction to provide the viewer with the kind of kinesthetic access to the process itself—not just the prediction. Transparency and physicality mattered as another way into the problem: they mattered as a form of lateral grasp, to understanding.

BATHTUB BLACK HOLES

In 1986, Kip Thorne and Richard Price wrote a book on the "membrane paradigm" in black hole physics—an expression borrowed from Thomas Kuhn's use of *paradigm* in his *Structure of Scientific Revolutions* of 1962.[37] An older view (paradigm) in astrophysics, they argued, held black holes as "collapsed stars," a notion captured in Russian by the phrase "frozen star." That view peered at the horizon from far away—and from that reference frame, indeed, phenomena at the horizon appeared to "hover for all eternity." But, Thorne and Price wrote, a newer picture had emerged since the mid-1960s, taking seriously the notion of a co-moving, infalling frame of reference ("the black hole's point of view") in which one followed a particle (or observer) plunging past the horizon toward the singularity within.[38] Seen from this frame, the term *frozen* seemed fully out of place.

Strictly speaking, it should not matter which point of view one took; in the fullness of calculation, they should yield the same result. But, in fact, it does matter. Not just here but throughout theoretical physics (the authors insisted), "a special role is played by the diagrams, pictures, mental images, and descriptive phrases that accompany our equations—pictures of magnetic field lines threading through a conducting plasma and the corresponding phrase that the field lines are 'frozen into the plasma.'" Indeed, Thorne and Price contended that "a new set of pictures and descriptive phrases can have a profound impact on the subsequent development of a field of research." Suddenly, the new way of thinking let one imagine the black hole horizon with an almost tangible physicality, as having conductivity, shear and viscosity, surface pressure, temperature, and entropy.[39]

Thorne's view was that we need to be able to go back and forth between these points of view—he was *not* arguing that the membrane paradigm was more or less fundamental than the differential geometry with which it was associated; there was no vertical hierarchy. Conceptual dexterity, as Thorne saw it, opened up new ways of solving problems. The analogy structure, as so often was the case, made possible problem solutions in both directions.

One of the relativists who contributed importantly to the exploration of the close-in dynamics near the horizon was William Unruh. Unruh, a relativist at the University of British Columbia, asked just how much you can learn from looking at the world—specifically black holes—through analogies and, importantly, with fluid analogies. In 1981, Unruh published a paper in which he bemoaned the fragility of Hawking's work on black hole evaporation—and the difficulty of testing it on what would need to be tiny black holes in the vicinity of Earth. "However," he continued, "a physical system exists which has all of the properties of a black hole as far as the quantum thermal radiation is concerned, but in which all of the basic physics is completely understood."[40] Here again, precisely the Maxwellian ambition: a set of mathematical relations realized in two domains, one in the radically curved space-time of a black hole and one in the easily explored domain of the physical (fluid) analog. Much could be gained in the analog system, maybe even the detection of quantum phenomena.

Unruh later put it this way: "Obviously there are things that you can learn about the world [through analogies]." One can, he says, "also learn a lot about it by seeing how independent the phenomenon is of those differences, so that if you have areas in which you're very *different* than the original thing that you're looking at, then if the effect really is independent of that difference, you get much, much stronger faith that the effect is really going to be there."[41]

For decades, many physicists considered black holes pathologies, as it were—unphysical or at least unrealized solutions to Einstein's eponymous equations. Then, in the early 1970s, astronomers detected material orbiting around an unseen gravitational source that indicated that the source might be so compact and so massive that it could not be made of any known kind of matter. In fact, there was no good explanation for what it could be other than a black hole. Black hole reality was reinforced at a much larger scale when stars could be seen dragged around the center of our galaxy—the Milky Way—in such a way and with such velocity that nothing but a supermassive black hole could be at work. More hints came from tremendous activity at the heart of many galaxies—perhaps these, too, were black holes, though much more massive. Then came LIGO's (Laser Interferometer

Gravitational-Wave Observatory, operated by Caltech and MIT) epochal gravity wave announcement of 2015—they had seen a signal of two black holes merging into a larger one. Bit by bit, these utterly unintuitive objects seemed to gain purchase on the physical world.

But how to understand them? Unruh often turned, surprisingly, if aptly, to Terry Pratchett's 1983 comic fantasy *The Color of Magic* where his imagined Discworld, among its many features, has an edge, a literal edge. Unruh, in a physics talk, found in that precipice an analog to the black hole: "to the fish, this Rimfall [the waterfall at the edge of the world disc] was a boundary, a horizon beyond which nothing could be heard. No fish who had ever travelled over the Rimfall had ever reported back. The shouts, or were they screams, of those intrepid explorers who had travelled too near to that boundary had suffered the most strange bass shifting, the fishes' high-pitched scream rapidly descending the scales to disappear from sound. Some claimed that with the most careful measurements of the sound, one could still hear sounds, of lower and lower frequency arbitrarily far into the future, as though the sound from those explorers never ceased. However, in no case could sounds ever be heard from beyond the location of that horizon, as that peculiar surface in the Rimfall came to be known."[42] So it was with black holes. A luminous object falling toward and eventually through the horizon would be seen (from far away) as shifting in color ever more toward the orange (lower frequency), with no light ever received from beyond the (black hole) horizon.

To get a better sense of the physics, Unruh began developing a fluid mechanical model of what he dubbed "dumb holes." The fluid mechanical action for the system can be rewritten to show the mathematical correspondence between sound waves in a flowing fluid and scalar fields on a background space-time. The terms in the metric depend on the fluid velocity such that when the radial velocity of the fluid is equal to the speed of sound, there is a "horizon" analogous to the black hole event horizon. Unruh went on to outline various uses of this mathematical similarity. The first is that many calculations made by general relativists may be applied to problems in fluid dynamics. "One may also obtain insight into black holes through fluid experiments," Unruh commented. "Since surface waves in an incompressible fluid look like scalar fields on a metric, we can do 'black hole physics in a bath-tub.'" The analog of the space-time metric, as it turned out, could be controlled by the variation in the depth of the tub, while an adjustable drain set the rate of inward flow. Now one could look at the direct analog—in water and drain—of a wave scattering from the bathtub vortex and picking up energy. This was precisely the analog of superradiance, in which waves gained energy when they hit the entrained movement of space itself around a black hole (the ergosphere).[43]

For both Maxwell and Lodge, electromagnetic phenomena could be demonstrated in the laboratory: their aim was to produce a parallel realization more accessible to the senses. Money could be tracked, but the machine could make manifest an analog process. But for these seeming sports of physics and astrophysics, the situation was different. Unruh: "We don't have little black holes around to . . . play with." But we do have whirlpools. "What [the whirlpool] illustrates is that the super radiant scattering is in fact very common. It's a ubiquitous phenomenon. It's not something that simply occurs for black holes, but occurs in everyday life. You get a big whirlpool in the ocean and there may well be situations in which the waves as they come in get amplified. What features are there of these kinds of systems that are going to produce super radiant scattering that are common to all of them."[44]

For theorist-turned-experimenter Silke Weinfurtner, there came a point when thought experiments, computer simulations, and analytic solutions were just not enough. In her words, the analog system of waves and whirlpools "is a completely different way of working. I'm a [theorist] by training, and . . . I thought I really understood that effect. I really understood it, I saw the derivations; I calculated myself. And then there's this real-life system and now everything has to be different. I have to rethink everything. How can I extract that?" So she did: after years of preparation and research, she came to run a "black hole laboratory" at the University of Nottingham, England. "There are . . . many layers of understanding something. And I always feel like the moment I've really understood an effect [is] when I've detected it." It "really changes the way you look at" the phenomena.[45]

In such work, Weinfurtner and Unruh joined the other figures we have discussed: none felt that the material models replaced the effects for which they are analogs; instead, the models dimensionalized understanding of the phenomena. As Weinfurtner puts it, "there are some beautiful moments in an experiment, right? When something works out, you understand something. And [if] one of these things . . . we wanted to understand [is there] you will see that; it is pretty nice. You send these plane waves—a wave front towards the vortex—and then you get this bizarre, beautiful pattern which you can see by eye. . . . And you say, 'What is it actually? Where is this coming from?' And then you have to sit down and say, 'Okay, which theory? What is the main thing?' Is [the] pattern . . . just super radiance because you take some of the waves out on one side and you amplify the others? Does that give a pattern? And it turned out the pattern itself is just that [of a] wave . . . moving on a curved space time geometry. It's like light bending. . . . It's geometry becoming visible" (Figure 2.9).[46]

Unruh underscored these fluid flow models. "If I could have a real black hole in my lab and I carried out this experiment [it would have been] very interest-

FIGURE 2.9. Bathtub black hole. Silke Weinfurtner's two-thousand-gallon "black hole laboratory" models many of the processes taking place just outside the horizon. Photograph by Peter Galison.

ing. . . . Carrying it out in another situation where we might think we understand the theoretical physics better than we do for black holes [as we do in the laboratory of surface waves and whirlpools] I think should have the same kind of weight as it would if one really carried it out for a black hole. It teaches us in fact even extra lessons for black holes. One could say okay, yeah, this happens for black holes. Who cares?" It seems rather unlikely that physicists might be so blasé, but the idea is clear: there is an underlying phenomenon captured by mathematical law and different ways of observing it, as in this diagram:

(black hole phenomena) ← (mathematical law) → (material analog)

Unruh, pressing the point, is saying that the law finds its realization and validation as much in an analog water wave setup as it does in light near a black hole. There is even the hope that the analog processes will lead to deeper waters, so to speak. One team, after contemplating the wide range of analog experiments, summed up their findings by saying that the analog physical models had been made to provide "very down-to-earth models of otherwise subtle behavior of general relativity." And yet there remained the speculative hope that "there may be more going on than just analogy—it is conceivable (though perhaps unlikely) that one or more of these analogue models could suggest a relatively simple . . . way of quantizing gravity that side-steps much of the technical machinery currently employed in such efforts." This faith in the thoroughgoing power of the liquid analogy to Hawking radiation was by no means undisputed. But enthusiasts went so far as to see in the analog and quantum phenomena different versions of the *same* universal phenomenon.[47] Opening up parallel presentations of the structures held the possibility that vortices as well as equations held different ways forward.

From Silke Weinfurtner's phrase "geometry becoming visible," we have in compact form the aim of these material models. Laws of science may find their expression in equations; they may find their predictive expression through simulations; they may find their experimental expression in the laboratory. But embodiment is something else again: these are the instruments of understanding.

UNDERSTANDING/STANDING AMONG

Pondering the different modes of reasoning about science (algebraic, geometrical, material), Maxwell advocated an epistemic openness: "for the sake of persons of these different types, scientific truth should be presented in different forms, and should be regarded as equally scientific whether it appears in the robust form and the vivid colouring of a physical illustration, or in the tenuity and paleness of a symbolical expression."[48] Acknowledging cognitive diversity has pedagogical consequences, as developmental psychologist Howard Gardner has insisted in a lifetime of work on "multiple intelligences."[49] Children, like adults, learn in different ways and retain more effectively when they approach a subject in more than one mode—with figures and texts, for example. No doubt material models play a role in such strategies of teaching and learning, offering another path to grasping what to some learners might be obscure in the purely algebraic-mathematical, for example. Our focus here, however, has been less on learning than on Maxwell's further claim, of greater import, that the modes he identified—symbolic, geometric, material—have equal claim to scientificity.

Indeed, Maxwell and our various other material modelers, from Oliver Lodge to Silke Weinfurtner, see phenomena manifested in different ways. What intrigues them is that something like superradiance can just as easily occur in the boosting of surface waves near a vortex as it can via light waves in the vicinity of spinning black holes. In both instances, waves can emerge with augmented energy. True, we might study the effect more felicitously in a two-thousand-gallon tank of dyed water than in the ergosphere of a black hole, but rightly interpreted, so say our various witnesses, we are observing the same underlying effect. Unruh labels this recognition as one of universalizing—which is one way of putting the lateral correspondence stressed here.

No doubt the concatenation of different realizations of a phenomenon bolsters learning, sharpens the intelligibility of the phenomenon. But it does something more important still: it gives us a greater variety of ways to carry on, alternate paths to extend the research to a next theoretical or experimental step. If electrodynamic, economic, or black hole material models truly succeed, they do so by offering other paths forward with different affordances.

Do these material, mathematical, and visually related analogies *explain*? In certain cases, they may. But the lateral correspondences are not, I would argue, principally aimed at explanation, the way a vertical (reductive or unifying) relation might be. Exaggerating for impact, Unruh noted that we might be working intensively on a phenomenon (like superradiance) manifested in fluid flow and yet be barely interested in actual black holes. His logic is clear: there is a phenomenon that is widespread (say, superradiance), manifested in different physical systems (around bathtub vortices and black holes). To this we can add that in such situations, it just cannot be that black holes *explain* vortical fluid flow *and* vortical fluid flow *explains* black holes. But it surely can be that focusing on the whirlpools may lead us to insights that might have been harder to apprehend through the study of differential equations governing space-time geometry.

Much of modern physical science is now advancing through such linkages, better captured by lateral correspondence than by vertical reduction. Indeed, in many engineering contexts, such a turn has grown increasingly clear. Modern system engineers regularly learn to link electrodynamic and mechanical domains: the analogies (and there are many) signal a commonality of structure. But they are not at all trying to *explain electrodynamics.* What matters for designers and engineers is the ability to carry on, and the lateral (analogical) structures allow them to do just that. In mechanical engineering, over the course of the twentieth century, it became commonplace to invert the Maxwellians project and to use, analogically, reasoning about electrical circuits and their components to design *mechanical* systems (like mechanical frequency filters).

In fact, the engineering sciences show us a stance that captures a certain indifference toward ontology and a studied focus on unifying structure. Consider Ilene Busch-Vishniac, author of one of the definitive turn-of-the-twenty-first-century treatises on the myriad devices, known as transducers, that take one form of energy and transform it to another.[50] Traditionally, transducers were thought to go from or to electrical domains: mechanical motion to/from electricity, heat to/from electricity, and so on. Her view is much less centered, no longer privileging voltages and currents. Instead, she sees transducers as embracing the full set of transformations in and out of chemical, electrical, thermal, mechanical, and other forms of energy. Some might try to partition optical, electrical, and magnetic processes; others prefer to group them together, because they all can be considered manifestations of Maxwell's theory. For Busch-Vishniac, such debates seem irrelevant. Instead, her interest is in the general set of phenomena captured by transduction: rotatory motion to electrical signals; magnetic fields to heat; indeed, the whole highly diverse set of transformations (every possible conversion between chemical, magnetic, thermal, optical, mechanical, and electrical), including transformations within a domain (e.g., magnetic to magnetic). To develop new sensors and actuators is to get to the business of calculating, designing, and realizing—with studied indifference to ontologies:

> While these philosophical debates are interesting, they are moot from the perspective of understanding transduction, because the focus is on conversion of energy in a device by such a means as to make it useful in monitoring a system or imposing a state on a system [sensing or transducing]. Indeed, the modeling technique that will be used in this book has been applied to many different energy domains, and to various manifestations of energy in the same broad domain with virtually no change required by shifting of the energy types.[51]

In designs of hearing aids, nail guns, vacuum cleaners, the back-and-forth between mechanics and electrical theory is constant. Analogies are tools to grasp relations and to build things. There are choices about how to set up these correspondences (between, say, springs and capacitors; dashpots [damped plungers] and resistors; and masses and inductors). But the goal is making.

Even if contemporary engineers extend the linkages of analogies far beyond where Maxwell took them, the nineteenth century, the spirit of Maxwell's analogizing holds fast. Indeed, in 1998, when Busch-Vishniac looked to ground her choice of the analogical structure, she chose the one called the Maxwell or impedance analogy (with mechanical, electrical, and fluid impedance forming a three-way correspondence). In the midst of a full range of late twentieth-century references, she cited one and only one much older work, James Clerk Maxwell's 1865 essay "The Dynamical Theory of the Electromagnetic Field."

If we wanted a slogan, it would be *correspondence without hierarchy*. Not a unification under a single set of entities or laws but instead an alignment of approaches, making common cause in understanding aspects of the world. When one is building so many forms of electromechanical, thermomechanical, or electro-optical devices, the lateral connections make reasoning vastly easier. Such a stress on lateral modeling has become ever more prevalent, and not just in engineering. One of the greatest contributions to theoretical physics in recent years is known as AdS/CFT, linking a form of curved space-time, anti–de Sitter space (an oft-used form of space-time geometry in gravitational theory), to conformal field theory (the language of particle physics). Some theorists have used AdS/CFT to extend insight *from* quantum field theory *into* gravity; others use the correspondence to extend insight *from* gravity *into* the quantum field theory of heavy ions (for example). Such cross-links expand our circle of understanding and mark much of the most intriguing tendencies in physics today.

There is an expanding form of unity within the physical sciences and engineering, but it is not the unity of laddered ascent (or descent from a single law or set of entities down through sciences considered derivative). Nor is that unity one of homogenization, just as electrodynamics does not *become* mechanics. Instead, distinct approaches with distinct ontologies find concordances—dualities, symmetries, corresponding structures. We are faced with a ring, linked profoundly in its parts but without a consensual center or peak: a ring, not a pyramid.

True, we live in an age when, to account for black holes, physics draws on abstract field theory, artificial intelligence, Monte Carlo simulations, topology, geometry, sensitive laser interferometers, and extended telescope arrays. And yet we can still work to find a way forward linking these directions—in the fineness of differential geometric abstraction and in the bulk of a two-thousand-gallon tank of swirling green water. Standing amid them all, tracing the lines of connection, perhaps we can find understanding.

NOTES

I would like to thank the editors of this volume for very helpful comments along the way and Erik Curiel, Jordi Cat, and members of the philosophy seminar within the Harvard Black Hole Initiative for precise and thoughtful suggestions. Appreciation, too, to Nicole Terrien for research and help with manuscript preparation. This project is funded in part by the Gordon and Betty Moore Foundation. It was also made possible through the support of a grant from the John Templeton Foundation. Opinions expressed in this publication are those of the author(s) and do not necessarily reflect the views of the John Templeton Foundation.

1 Catherine Elgin argues that the "felicitous falsehoods" of such idealizations exemplify facts, even though the idealizations are, strictly speaking, false—and that this process of exemplification delivers a form of understanding. Elgin, "Understanding and the Facts," *Philosophical Studies* 132 (2007): 33–42.

2 There is a quasi-infinite literature on modeling in the sciences, but a recent collection offers a good launching point with some superb essays addressing periods from the eighteenth century forward: Soraya de Chadarevian and Nick Hopwood, eds., *Models: The Third Dimension of Science* (Stanford, Calif.: Stanford University Press, 2004).

3 Jonathan Kvanvig takes understanding to be an "honorific"—that is, understanding serves as nothing more than a compliment, not a justified epistemological state of affairs. See Kvanvig, "Knowledge and Understanding," in *The Value of Knowledge and the Pursuit of Understanding,* 185–203 (Cambridge: Cambridge University Press, 2003). This demotion of the concept of understanding has a long history in recent decades. Carl Hempel, in *Aspects of Scientific Explanation and Other Essays in the Philosophy of Science* (New York: Free Press, 1965), 413, clearly considered understanding to be nothing more than a psychological trait, not part of the knowledge structure itself. That line of assessment has continued, as Henk W. de Regt and Dennis Dieks rightly contend, in work by Bas van Fraassen and J. D. Trout—see their "A Contextual Approach to Scientific Understanding," *Synthèse* 144 (2005): 137–70, esp. 141.

4 Michael Friedman, "Explanation and Scientific Understanding," *Journal of Philosophy* 71 (1974): 18.

5 Henk W. de Regt, *Understanding Scientific Understanding* (Oxford: Oxford University Press, 2017), 12. My emphasis here is more on understanding phenomena than theories per se. I very much appreciate Howard Stein's formulation of understanding as fundamentally joined to the clarification of concepts. "The Enterprise of Understanding and the Enterprise of Knowledge," *Synthèse* 140 (2004): 135–76.

6 Jordi Cat, "On Understanding: Maxwell on the Methods of Illustration and Sci-

entific Metaphor," *Studies in History and Philosophy of Science, Part B: Studies in History and Philosophy of Modern Physics* 32, no. 3 (2001): 436.

7 On forms of coordination, trading zones, and scientific interlanguages, see Galison, *Image and Logic: A Material Culture of Microphysics* (Chicago: University of Chicago Press, 1997).

8 *Online Etymology Dictionary*, s.v. "understand," https://www.etymonline.com/word/understand.

9 For more on analog models across a wide range of fields, see Susan G. Sterrett, "Experimentation on Analogue Models," chapter 39 in *Springer Handbook of Model-Based Science*, ed. Lorenzo Magnani and Tommaso Bertolotti, (New York: Springer, 2017), and further references therein. Sterrett rightly points out how widespread the analog models are and have been—and that digital simulations often must contain vastly more information than "similarity models."

10 On the vertical (pyramid) versus lateral (ring) forms of linkages among theories, see Peter Galison, "The Pyramid and the Ring: A Physics Indifferent to Ontology," in *Research Objects in Their Technological Setting*, ed. Bernadette Bensaude Vincent, Sacha Loeve, Alfred Nordmann, and Astrid Schwarz, 15–26 (New York: Routledge, 2017), and his "Meanings of Scientific Unity: The Law, the Orchestra, the Pyramid, Quilt, and Ring," in *Pursuing the Unity of Science: Ideology and Scientific Practice from the Great War to the Cold War*, ed. Harmke Kamminga and Geert Somsen, 12–29 (Burlington, Vt.: Ashgate, 2016).

11 James Clerk Maxwell, "A Dynamical Theory of the Electromagnetic Field, 1864," in *The Scientific Papers of James Clerk Maxwell*, ed. W. D. Niven (New York: Dover, 1965), 1:563–64. It is exactly this contrast between understanding and explanation that I find intriguing, and not because understanding it is *merely* illustrative; not at all—instead, understanding emerges from the complementarity of different realizations of a theory. Emphasis on "understanding" added in the Maxwell quotation.

12 James Clerk Maxwell, "Address to the Mathematical and Physical Sections of the British Association," in Niven, *Scientific Papers*, 2:219, hereinafter Maxwell, "Presidential Address." The literature on models in Maxwell and Maxwellian electrodynamics is vast. Intrinsically outstanding and of use for further reference are, for example, Jed Z. Buchwald, *From Maxwell to Microphysics: Aspects of Electromagnetic Theory in the Last Quarter of the Nineteenth Century* (Chicago: University of Chicago Press, 1985), which makes clear that the ether models became a huge difficulty later when faced with Lorentz's work toward a particle and field picture of microphysics, and Daniel Siegel, "The Origin of the Displacement Current," *Historical Studies in the Physical Sciences* 17 (1986), 99–146; for a very perspicuous synthetic work on electrodynamics including the role of models for Maxwell in his electrodynamics, see Olivier Darrigol, *Electrodynamics from Ampère to Einstein* (Oxford: Oxford University Press, 2000), esp. 147–63. See

also Cat, "On Understanding." He contextualizes many of Maxwell's views in the philosophical psychology of Victorian England and stresses, rightly in my view, that the kind of understanding he sees as a goal of Maxwell's work is not fundamentally tied to Maxwell's idea of explanation.

13 Maxwell, "Presidential Address," 219–20.

14 Maxwell, 220.

15 Maxwell, 220.

16 James Clerk Maxwell, "On a Dynamical Top, for exhibiting the phenomena of the motions of a body of invariable form about a fixed point, with some suggestions as to the Earth's motion," in Niven, *Scientific Papers of James Clerk Maxwell,* 1:248.

17 Sigmund Freud, "Creative Writers and Day-Dreaming," *Standard Edition* 9 (1959): 142–53, "date-mark" on 147.

18 Maxwell, "On Potential," in *The Scientific Letters and Papers of James Clerk Maxwell,* ed. P. M. Harman (Cambridge: Cambridge University Press, 1990), 3:787, emphasis added.

19 Maxwell, "Presidential Address," 219. Maxwell more extensively treats the idea of classification in his 1869 "On the Mathematical Classification of Physical Quantities," in *The Scientific Papers of James Clerk Maxwell,* ed. W. D. Niven (New York: Dover, 1965): 257–66. Here Maxwell distinguishes the formal classification from the physical classification. Thanks to Erik Curiel for discussion of this paper.

20 Maxwell, 226–27.

21 Maxwell, 227.

22 Maxwell, 219. Bruce Hunt on "broadening" (with which I completely agree), *The Maxwellians* (New York: Cornell University Press, 1994), 74, though presenting this as broadening the paths to "truth" may be overstated.

23 Oliver Lodge, *Modern Views of Electricity* (London: Macmillan, 1907), 12–13.

24 Lodge, 212. See also Hunt, *Maxwellians,* 89–93, for an excellent discussion of this and similar mechanical model figures.

25 Lodge, *Modern Views,* 60–61. See also Hunt, *Maxwellians,* 90–91.

26 Lodge, *Modern Views,* 60.

27 Oliver Lodge, *The Ether of Space* (New York: Harper, 1909), xviii.

28 Alan Bollard, *A Few Hares to Chase: The Economic Life and Times of Bill Phillips* (Oxford: Oxford University Press, 2006), 28ff.

29 Phillips wrote about this to Richard Sayers, in Bollard, *A Few Hares to Chase,* 98–99; Mary Morgan, *The World in the Model: How Economists Work and Think* (Cambridge: Cambridge University Press, 2012), esp. chapter 5 (172–216), where she addresses the Phillips–Newlyn Machine with great clarity, justly evoking Newlyn's contribution in much greater depth than I can here.

30 Bollard, *A Few Hares to Chase,* 101.

31 Nicholas Barr, "The History of the Phillips Machine," in *A. W. H. Phillips: Collected Works in Contemporary Perspective,* ed. Robert Leeson (Cambridge: Cambridge University Press, 2000), 92.

32 Nicholas Barr, "The Phillips Machine," *LSE Quarterly* 2 (1988): 310, discussion of Robertson and Keynes on 329.

33 Morgan, *World in the Model,* 210. A good summary of the dispute can be found in Andres Nentjes, "The Keynes versus Robertson Controversy in Monetary Economics with a Post-Scriptum on the Dutch Monetary Controversy in the Nineteen Fifties," *De Economist* 127, no. 4 (1979): 551–69.

34 A. W. H. Phillips, "Mechanical Models in Economic Dynamics," *Economica,* New Series 17, no. 67 (August 1950): 283.

35 Lionel Robbins recalling the demonstration from November 1949, quoted in Bollard, *A Few Hares to Chase,* 108.

36 Phillips, "Mechanical Models in Economic Dynamics," 283.

37 Thomas Kuhn, *The Structure of Scientific Revolutions* (Chicago: University of Chicago Press, 1962).

38 Richard Price and Kip S. Thorne, introduction to *Black Holes: The Membrane Paradigm* (New Haven, Conn.: Yale University Press, 1986), 1.

39 Price and Thorne, 1. The geometrical approach advocated by Price and Thorne draws deeply on the geometrical (geometrodynamical) approach advocated by Charles Misner, Kip Thorne, and John Wheeler in their generation-marking textbook *Gravitation* (Princeton, N.J.: Princeton University Press, 1973).

40 Unruh goes on to sketch the lines of a possible inquiry: "In this system one can investigate the effect of the reaction of the quantum field on its own mode of propagation, one can see what the implications are of the breakdown of the wave equation at small scales on the evaporation process, and one might even contemplate the experimental investigation of the thermal emission process." William G. Unruh, "Experimental Black-Hole Evaporation?," *Physical Review Letters* 46 (1981): 1351.

41 William G. Unruh, interview with Peter Galison, Cambridge, Mass., April 23, 2018, hereinafter, Unruh and Galison, interview.

42 William G. Unruh and Ralf Schützhold, eds., *Quantum Analogues: From Phase Transitions to Black Holes and Cosmology* (Berlin: Springer, 2007), 1.

43 Matthew Hasselfield, "A Graduate Student Summary of the Public Lecture 'Black Holes/Dumb Holes: Condensed Matter Analogues' by Bill Unruh," PITP Showcase Conference, May 2005, https://pitp.phas.ubc.ca/archives/CWSS/showcase/unruhsum.html. "Hawking radiation seems to give much insight into black hole physics, in particular by providing a framework for black hole thermodynamics. The thermodynamical justification for Hawking radiation is so strong that Unruh feels that 'this thermodynamic analogy makes one believe that the result is right even though the derivation is nonsense.'"

44 Unruh and Galison, interview.

45 Silke Weinfurtner, interview with Peter Galison, Cambridge, Mass., May 9, 2018, hereinafter Weinfurtner and Galison, interview.

46 Weinfurtner and Galison, interview. Also a strong counterargument to the force of

analogue experiments on Hawking radiation: Karen Crowther, Niels Linnemann, and Christian Wuthrich, "What We Cannot Learn from Analogue Experiments," *S.I.: Reasoning in Physics,* 2019, https://link.springer.com/article/10.1007/s11229-019-02190-0. My own view is that the analog experiments work well in addressing superradiance, which is not "as inaccessible" (in several ways) as Hawking radiation.

47 Carlos Barceló, Stefano Liberati, and Matt Visser, "Analogue Gravity," ArXiv:gr-qc/0505065v3 (May 12, 2011), 108. For a discussion of the philosophical views in favor of and against the analog models of Hawking radiation, see Erik Curiel, "Singularities and Black Holes," in *The Stanford Encyclopedia of Philosophy,* Spring 2021 ed., ed. Edward N. Zalta, https://plato.stanford.edu/archives/spr2021/entries/spacetime-singularities/.

48 Maxwell, "Presidential Address," 220.

49 Gardner's first major presentation of these ideas was in Howard Gardner, *Frames of Mind: The Theory of Multiple Intelligences* (New York: Basic Books, 1983) and then was developed and debated in many contexts among educators and psychologists.

50 Ilene J. Busch-Vishniac, *Electromechanical Actuators and Sensors* (New York: Springer, 1998), 2–4.

51 Busch-Vishniac, 5.

Part II

Sensing

3

William Farish's Devices and Drawings: Models for Envisioning Immaterial and Material Realms

HILARY BRYON

In 1795, and barely a year into his tenure as a Cambridge University professor and natural philosopher, William Farish began using three-dimensional modeling mechanisms to make physiomathematic theories present to observers. His modeling system consisted of a kit-of-parts that would be assembled to demonstrate the mechanics behind various machines. The apparatus comprised a provisional scaffold of wood to which various metal mechanical elements, such as wheels, axles, and pulleys, would be attached via clamps (Figure 3.1). As the model system would be disassembled on a regular basis so its various parts could be reused for new demonstrations, Farish formulated a drawing technique, called "isometrical perspective." While not technically a perspective, the nomenclature indicated a pictorial form of representation. The drawings not only documented the exact construct of the temporary, individual assemblies but allowed for the successive replication of their material construction.[1] In this way, Farish was employing both analogic and iconic models, the first being models of immaterial principles made visible through a modeling device, the second being models of material artifacts made visible through a mode of drawn, pictorial representation that is uniquely object related. Analogic modeling necessarily involves a move toward abstraction, and Farish's ephemeral analogic models communicate a conceptual understanding; his material modus of modeling, a kit-of-parts, supports an imaginative flexibility toward demonstrating immaterial, physiographic principles. In turn, Farish's isometric drawing system was designed to record these physical, yet temporary, analogic modeling devices. These drawings thus operate as iconic models. Indeed, unlike other pictorial modes of two-dimensional representation, such as perspective or oblique projection, isometric projection offers an unparalleled, isomorphic manifestation of a three-dimensional material referent through the drawings'

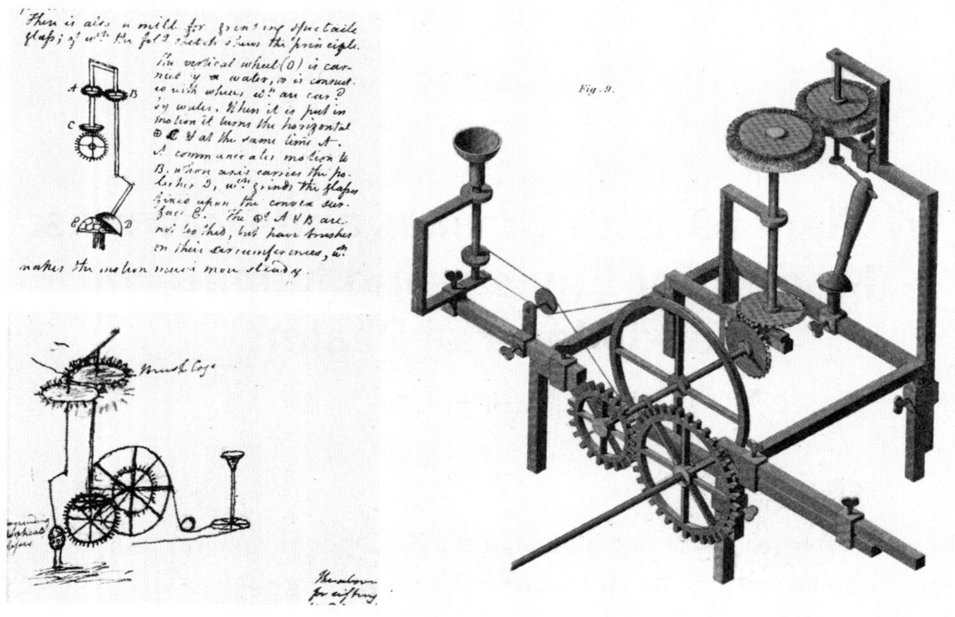

FIGURE 3.1. *Left,* crude, two-dimensional depictions in student sketches of a mechanism for grinding optical glass interleaved within copies of Farish's *Plan of a Course of Lectures on Arts and Manufactures* pamphlets (*above,* 1803; *below,* 1796). *Right,* drawing of a model system assembled to reveal the same mechanism. Farish's illustration iconically models his kit-of-parts modeling system as well as his isometric, parallel projection drawing system. William Farish, "On Isometrical Perspective," *Transactions of the Cambridge Philosophical Society* (1822), Figure 9. Photographs by the author.

object-oriented spatial sensibility. This essay will contextualize these modeling practices relative to antecedent and subsequent developments so as to establish their significance in negotiating the realms of representation tied to devices and drawings, the immaterial and material, and abstraction and mimesis. Farish's history also reminds us that models model their own representational strategies as well as specific referents.

Architectural educator and philosopher Salahuddin Choudhury offers insight into iconic and analogic models that helps us distinguish the dynamic of transition inherent to the kind of work models can do. Iconic models, according to Choudhury, "operate by virtue of a *visual isomorphism* between the event and the model. Accordingly, the visual structure—the critical relations of parts as perceived—is kept invariant."[2] Analogic models, on the other hand, "operate by virtue of a *logical isomorphism* between the event and the model. Accordingly, the logical structure—the critical relations of elements as cognized—is kept invariant."[3] Representations of an iconic approach to modeling are scaled likenesses aimed at accurately resembling the referent, whereas analogic forms of modeling strive to convey a conceptual structure that critically distills and exposes a system, situation, or process. Thus, an iconic model establishes a proportional image relating literally and mimetically to the physical, material world, whereas an analogic model maintains a focused functional likeness or structural correspondence.

William Farish's use of both forms provides a compelling portrait of models as curiously promiscuous representations that reach into imagined, creative realms even as they purport to bring about a more faithful, scientific understanding of the world they model. The context also offers a sketch of how model thinking and model-making resonate in different disciplines. In what follows, I discuss the context, precedents, and significance of William Farish's analogic modeling system to consider how the novel drawing system he developed is uniquely object related and so an iconic model. Both modes of representation that Farish piloted translate between the material and the immaterial. The two modes are, to some degree, inversions of each other. The analogic, kit-of-parts model makes invisible forces fleetingly visible, whereas the iconic isometric parallel projection drawing makes an object visible through an immaterial depiction that parallels the material, spatial reality and measured accuracy of the object itself. Ultimately, Farish's models reveal how seemingly small design articulations shift *what* is being communicated through *how* it is being communicated; his historical context testifies to the way modeling assumes different imaginative forms.

MODELING THE IMMATERIAL: NATURAL PHILOSOPHY, THE SENSES, AND KINEMATICS

From antiquity to the nineteenth century, the study of natural bodies and the phenomena connected with them was known as natural philosophy.[4] In this field, Isaac Newton (1643–1727), one of Farish's predecessors at the University of Cambridge, stands out as the perhaps most notable of natural philosophers, earning part of this acclaim through the publication of his *Philosophiae Naturalis Principia Mathematica* (*Mathematical Principles of Natural Philosophy*, 1687). Developed primarily while in residence at Cambridge as the Lucasian Professor of Mathematics (1669–1702), Newton's *Principia* set out the foundations of modern mechanics and physics through his mathematical determination of laws of motion and gravitation.[5] These laws explained observable and related natural phenomena. However, for many readers, Newton's principles were difficult to understand, given the abstract nature of mathematical proofs and the fleeting, immaterial forms of the phenomena under description. Thus, in the early eighteenth century, one is not surprised to see how the teachings of Newton's philosophy, specifically those principles related to mechanics, turn toward the demonstration of various physical effects with three-dimensional models. A little over a decade after *Principia* was published, practical demonstration lectures devoted to the fundamental principles governing mechanics began at Oxford University in 1700 and at Cambridge University in 1707.[6]

As these practical displays became part of university instruction in the early eighteenth century, the Dutch natural philosopher Willem Jacob 's Gravesande (1688–1742) argued that models—and the way they engaged the senses—were essential for perceiving the immaterial truths of the natural world. In 1715, he visited England, met Isaac Newton, and, two years later, became a professor of mathematics and astronomy at Leiden University in the Netherlands.[7] 's Gravesande's course, conveyed in his book *Mathematical Elements of Natural Philosophy, Confirmed by Experiments; or, An Introduction to Sir Isaac Newton's Philosophy* (1724), includes "An Oration Concerning Evidence."[8] Within, 's Gravesande states that the senses, testimony, and analogy are the "foundations of persuasion" and truth. He recommends that we should make use of our own senses and the observations of others and use the "power of applying our observations to things not observed."[9] In a book ostensibly devoted to mathematical principles, 's Gravesande establishes sense perception and analogic modeling as critical poles in understanding and transmitting the fleeting presence of mechanical principles.

A century later, analogic sensibility continued as a critical guiding principle in the articulation of kinematics, or the study of movement. In 1838, André-Marie Ampère (1775–1839), French physicist and mathematician, classified the study of mechanisms as a unique branch of knowledge "in which movements are considered in themselves [independent of the forces which produce them], as we observe them in solid bodies all about us, and especially in the assemblages called machines."[10] Looking to the Greek word for "movement" to identify this science of movement, mechanism, and elementary mechanics, Ampère settled on the term *kinematics* to describe this theoretical model.[11] Three years later, Robert Willis, Cambridge University professor and successor to William Farish, contributed to the standardization of the term *kinematics* through its use within his widely distributed and acclaimed book *Principles of Mechanism*.[12] Willis noted in this volume that while publications on the subject matter of specific machines or machinery can be traced back to the fifteenth century, it was not until the eighteenth century that a sense of kinematics entered the scientific discourse. Willis, citing the eighteenth-century mathematician Leonhard Euler, advocated that "the investigation of the motion of a rigid body may be conveniently separated into two parts, the one geometrical [statics], the other mechanical. . . . Mechanics, the determination of motion from dynamical principles, will be made much easier than if the two parts were undertaken conjointly."[13] This discourse demonstrates that studying movement itself was separate from studying the forces that produce movement. Ultimately, the conceptual systems of motion and mechanics were distilled and evidenced through analogic modeling devices by people like Farish. These analogic models, as we will see, had the pedagogic goal to model kinematic principles, not the machines utilizing them.

SENSING MOVEMENT: KINEMATIC MODELING DEVICES (FROM PHILOSOPHICAL TABLE TO PROTEAN MECHANISM)

The first demonstration models used by natural philosophers were used to illustrate applied mechanics.[14] Often, these models were either discrete, fixed assemblies devoted to isolated components or literal examples of machines. The former became codified with titles such as the "lever," the "wheel and axle," "pulleys," or "screw and worm," and the latter evidenced by functional, scaled representations of actual machines, such as steam engines, catapults, or pile drivers. The fixed-component assemblies, as hybrid analogic–iconic models, helped demonstrate the mechanical advantages of variable adjustments within a component set, such as changing the position of a fulcrum relative to a lever or evaluating the different turning ratios between wheels with different diameters. The iconic models, in contrast, demonstrated the whole machine as a singular, fixed object. Together, though, these models explicated the pragmatic functionality of components—they helped make visible the forces at work in mechanical devices of differing scales through analogic as well as iconic means (Figure 3.2).

A less conventional, more analogic approach toward modeling emerged in relation to kinematic principles in the mid-eighteenth century: the composite apparatus. Composite apparatuses operated as analogic modeling devices, isolating distinct kinematic phenomena to make immaterial principles visible. One of the earliest examples of this approach was initiated by Willem Jacob 's Gravesande. It was first designed by his Dutch instrument maker Johannes Joosten van Musschenbroek and further standardized by London-based instrument maker to King George III, George Adams.[15] Within the pages of 's Gravesande's textbook, one finds descriptions and illustrations of the basic modeling components for what was later simplified and identified by George Adams as a *Philosophical Table* (Figure 3.2).[16] The critical distinguishing elements include a wooden base table to which one would affix wooden scaffolding elements, such as vertical boards or pillars and horizontal arms, as well as connecting elements, such as copper screws, plates, hooks, couplings, rules, and angles, plus rigging threads, strings, and ropes. The various configurations, referred to as "machines" by 's Gravesande, were assembled per the typology of experiments being observed, one such being "a machine whereby many experiments, of innate forces and the collision of bodies, are made."[17] A given

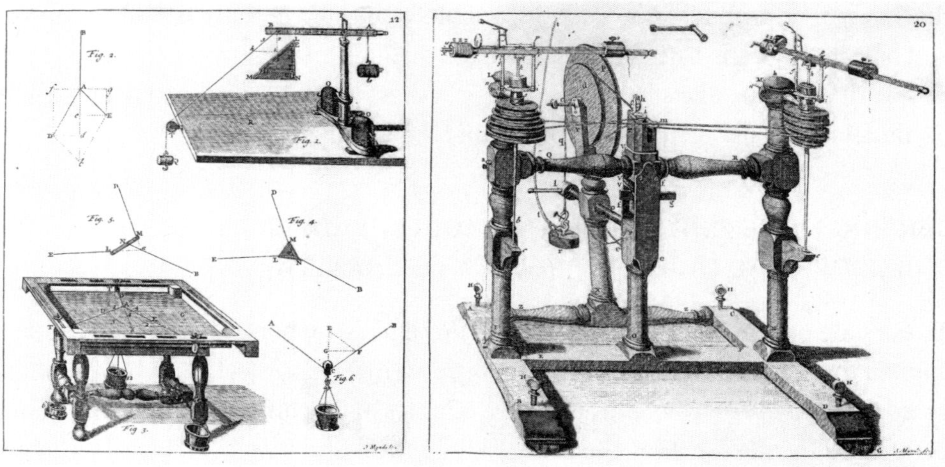

FIGURE 3.2. *Left,* a board with pillar, pulleys, strings, and so on to be affixed atop the base of Philosophical Table for experiments with "Oblique Powers." *Right,* wooden frame and attachments for experiments with "Central Forces." From 's Gravesande, *Mathematical Elements,* 6th ed. (1747), Plates 12 and 20. Photographs by the author.

model assembly would make a system of kinematic phenomena manifest. With the standardization brought by George Adams, 's Gravesande's modeling marks one of the earliest turns toward a more system-based, analogic approach. It shows the turn toward more open-ended models that could relay a set of kinematic principles, rather than model a single machine's mechanical workings. The flexibility of the modeling device, rather than a fixed model, is what came to matter. The composite apparatus, as a model itself, then illustrates the related but changing properties of kinematics, so that modeling itself was changed by the content it sought to model.

In the late eighteenth century, such composite modeling apparatus for kinematic demonstrations were transformed at the University of Cambridge under William Farish, who was professor of chemistry from 1794 to 1813 and Jacksonian Professor of Natural and Experimental Philosophy from 1813 until his death in 1837. Spanning these two professorships, Farish's lectures were oriented toward natural philosophy, from material matter to immaterial mechanisms. His 1796 "Plan of a Course of Lectures on Arts and Manufactures, More Particularly Such as Relate to Chemistry" consisted of four parts: "Metals and Minerals," "Animal and Vegetable Substances," "On the Construction of Machines," and "Water-works and Navigation." Farish would not publish information on his modeling system until 1822, but student sketches testify to Farish's use of demonstration models in his lectures (Figure 3.1). Upon Farish's death, Robert Willis (1800–75) succeeded him as Jacksonian Professor of Natural Philosophy, holding the post from 1837 to 1875. Willis carried on Farish's traditions in more ways than one. Like Farish, Willis

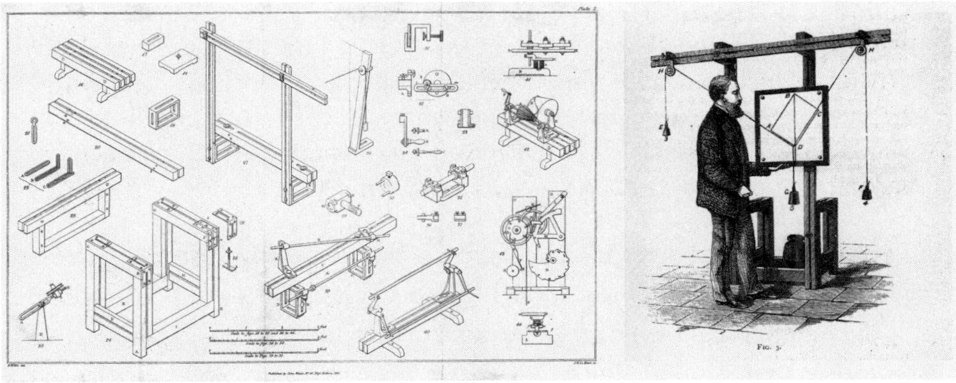

FIGURE 3.3. *Left,* Robert Willis, arrayed components of the *System of Apparatus for the use of Lecturers and Experiments in Mechanical Philosophy* (1851), Plate 2, depicted through Farish's isometric parallel projection drawing system. *Right,* Willis's apparatus adopted by Ball to illustrate his own lectures and depicted through perspectival projection and in use, from *Experimental Mechanics* (1888), Figure 3. Photographs by the author.

lectured on mechanical philosophy, but furthermore, he "adopted Farish's method of illustrating his lectures by means of models built up of component parts."[18] Unlike Farish, Willis published a detailed account of the modeling system. *The System of Apparatus for the Use of Lecturers and Experimenters in Mechanical Philosophy* (1851) communicated their shared teaching method as well as Willis's more refined construction kit used to demonstrate the mechanical principles underpinning a large array of machines (Figure 3.3, left). Many adopted the Farish–Willis system of apparatus, most notably famed astronomer and mathematician Sir Robert Stawell Ball (1840–1913). Ball lectured on experimental mechanics at the Royal College of Science in Dublin.[19] In 1871, Ball published his course of lectures and demonstrated the modeling apparatus in use (Figure 3.3, right). Ball's perspectival illustrations envision for us how mechanical principles were made visible during the act of demonstration.

Farish's obituary also described the context and import of his pedagogic, analogic modeling as follows:

> Having taken a personal survey of all the districts in Great Britain, in which mining, manufactures, or agriculture, were especially cultivated, [William Farish] provided himself with elements for the construction of *models on a new and most ingenious plan.* By means so simple that they excited no wonder, he presented working models of almost every machine which he described, and certainly every fundamental form of mechanical contrivance. A few bars, axles, clamps, wheels, etc. appeared in different forms throughout the lectures, in all the diversified constructions which it was necessary to explain.

> Nothing could be better adapted for the purposes of University instruction: the sameness of the elements employed excluded all the peculiarities of structure which depend upon particular details, and which are frequently the most striking external parts of a machine. In the lectures of Professor Farish, *identity of principle was remembered by identity of structure, whether the contrivance were part of a watch or a watermill.*[20]

Thus Farish's audiences came to understand the intended distinction between modeling a mechanical principle and modeling a machine. His "identity of structure" fulfills the requirement of logical isomorphism demanded of analogic models. As his protean system of modeling materialized mechanical principles, such as vectors or momentum, it aided students to visualize, or sense, the immaterial realm. Conversely, by assembling a modeling device of component parts, the material model becomes, in a sense, abstract.

ISOMETRIC "PERSPECTIVE"

Farish's novel modeling system led him to codify a novel drawing system.[21] This form of orthographic parallel projection drawing—isometry—became, and remains today, a key representational strategy. "On Isometrical Perspective" was first presented to the Cambridge Philosophical Society in 1820 and then published in its *Transactions* in 1822. Farish explained the need and the struggle to clearly document his temporary modeling devices so that they could be stored and reassembled by assistants at future dates. Examining the representational struggles encountered in the sketch notes drawn by students, particularly with respect to spatial depth cues, Farish's challenge is clear (Figure 3.1). He determined that conventional three-view drawings—juxtaposed plans, sections, and elevations—recorded an assembly's scaled, material dimensions with precision, but the aggregated set of individual two-dimensional orthographic drawings does not allow an apparatus to be understood as a three-dimensional entity in space.[22] Conversely, Farish observed that "common perspective" could be used to render a picture but that a conic projection significantly distorts the material reality of the object since its parallel lines were not represented as parallel and the representation has no measurable scale. Thus Farish's "isometrical perspective" combined attributes of both parallel and perspectival projection systems, and so one image uniquely united the visual realm of appearance with the physical dimensions of the object (Figure 3.4).[23]

In presenting his novel projection drawing, Farish acknowledged that although "isometrical perspective" is not actually a perspective, it does provide a pictorial view.[24] A visual isomorphism is achieved between the object and its pictorial

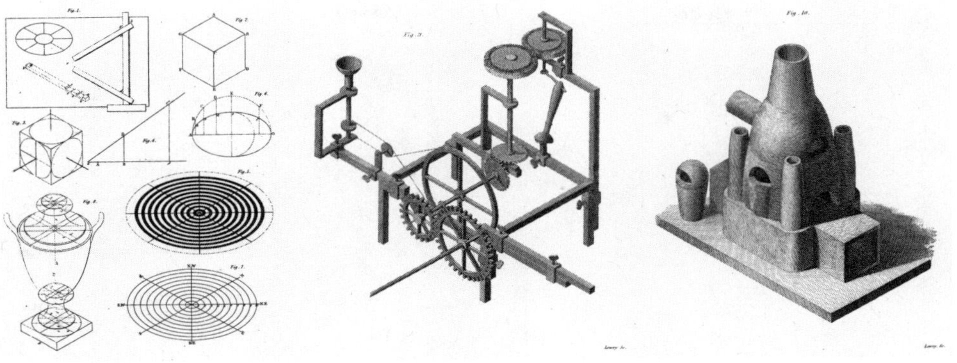

FIGURE 3.4. Farish's "On Isometrical Perspective," *Transactions of the Cambridge Philosophical Society* 1 (1822), Figures 1–10. The three plates relate, *left to right,* the geometrical and practical construction of the drawing system, a protean–kinematic model assembly recorded through isometry, and a furnace rendered through an isometric. The fixed isometrical parallel projection drawings record a representation in equal measure to the object in space. Photographs by the author.

representation that indicates the drawing's affinity with iconic models. The isometric "perspective" creates a picture by placing the eye of the observer at an indefinite distance, rather than at a fixed point.[25] And being an orthographic parallel projection, in which a three-dimensional object is portrayed by parallel projectors perpendicular to the drawing surface, it is scaled. Farish illustrated the approach through an isometric drawing of a cube (Figure 3.5), and one can easily perceive its three principal faces equally inclined to the picture plane and so demonstrating the drawing's name: "isometric," which literally means "equal measure."[26] The critical elements of what is later referred to as isometry, isometric drawing, or isometric projection are plainly established by Farish: measurable scale and a spatial depiction relationally equal to the object being referenced. Isometric drawing was advanced by many, including early twentieth-century architect Claude Bragdon, who noted that it is "less faithful to appearance, (and) more faithful to fact; it shows things more nearly as they are known to the mind: Parallel lines are really parallel; there is no far and no near, the size of everything remains constant because all things are represented as being the same distance away and the eye of the spectator everywhere at once. When we imagine a thing, or strive to visualize it in the mind or memory, we do it in this way, without the distortions of ordinary perspective."[27] Farish's isometric parallel projection drawing indicates an absolute relation to the object's material, spatial reality and thus a unique capacity to itself operate isomorphically as a physical iconic model. Nevertheless, an inherent tension between isometric projection and perspectival perception remained, and in being contested, "isometrical perspective" evolved.

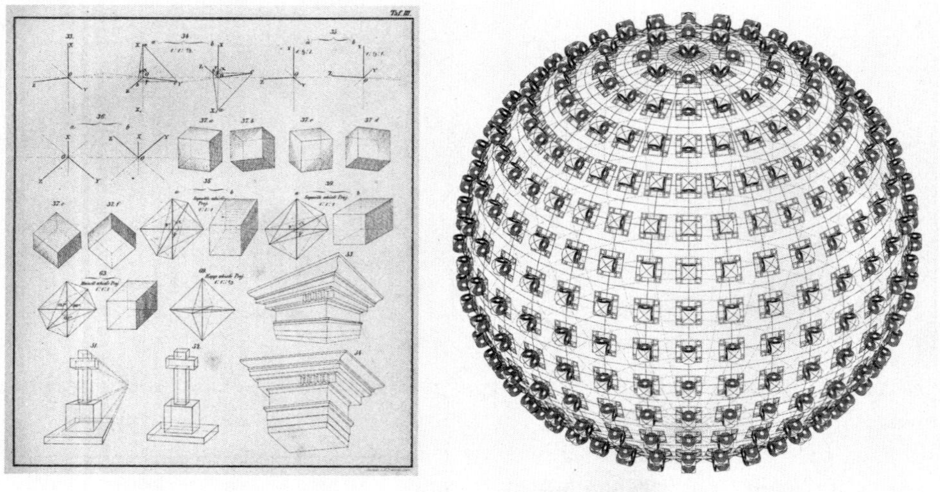

FIGURE 3.5. *Left,* Meyers's Plate III: selection of an axonometric axis system establishing the orientation of the inclined object in space and so a particular viewing angle. Figures 51–52 (crosses) and 53–54 (entablatures) juxtapose perspective projections to dimetric parallel projections to convey perceptual correspondence in *Lehrbuch der Axonometrie* (in series 1853–55). *Right,* Patrice Boissy's drawing "La Sphère" demonstrating an array of axonometric views and thus establishes axonometry's capacity to fix a "good angle." The isometric projection centers the projection sphere in the uppermost zenith position. Jean Aubert, *Axonométrie, théorie, art et pratique des perspectives parallèles* (Paris: Éditions de la Villette, 1996), 97. Image courtesy of École Nationale Supérieure d'Architecture de Paris-La Villette. Photographs by the author.

AXONOMETRIC POINTS OF VIEW

In the years following the publication of Farish's "isometrical perspective," its methods and application were debated, advanced, and developed by successive authors.[28] In due course, isometry was significantly reconfigured and subsumed as a class of axonometry.[29] The first, and most comprehensive, discernment and advancement of axonometry as a form of representation was achieved by mathematicians M. H. Meyer and C. Th. Meyer. In their *Lehrbuch der axonometrischen Projectionslehre* (1855–63), the brothers examined the history of parallel projection so as to support their development of a new system of pictorial parallel projection: axonometric. They singled out the influential work of their compatriot Julius Weisbach.[30] Weisbach proposed that isometric orthographic projection, which is fixed and uniform as embodied by the equally inclined cube, could be reconceived as a special case of a spatial three-dimensional axis system.[31] Weisbach theorized that points could be projected by their coordinates on axial planes in space. As such, the orientation and reduction of the axes would not have to stay isometrically inclined and equally reduced.[32] As intimated by Weisbach, axis rotation could allow a parallel projection to establish a more natural relation between an object and its representation to a viewer.[33]

Meyer and Meyer proposed that the axis system observed by Weisbach, comprising isometric, monodimetric, and anisometric projections, be named "axonometry" and that the method be called "axonometric projection."[34] The Meyer brothers recommended that axonometric projection drawings should begin by first determining the desired orientation of the axonometric axes and by consequence the orientation of the inclined object in space (Figure 3.5, left).[35] To establish the orientation of the axis system, the object is imagined as to how it naturally falls into the picture, or projection, plane. The brothers asserted that maintaining a natural aspect will more closely approximate the image as one receives it to the eye and thus satisfy what they consider to be the first prerequisite for image making.[36] This natural relation of the object to the eye is not considered in isometric projection due to its completely systematic, objectifying application. The Meyers believed that an object was best represented by axonometric projection because it entails a closer correspondence between the observer and the object through the discrete orientation of the axis system. Thus the selection of the axonometric angles is a critical representational choice, as their selection can reflect, heighten, or diminish a viewer's perception of particular aspects of a given object. In a way, the development of axonometry upends the objectifying isomorphism that is implicit to Farish's isometric projection. At the same time, it underscores the virtues of the iconic model. Because axonometry's visual structure is tied to a viewer's orienting perception, it is less tied to the object. Inversely, axonometric projections suggest a mode of graphic modeling in which the kinetic aspect is the "point of view," and so axonometry makes available aspects of representation previously inaccessible in parallel projection. Moreover, as a technique of observation, axonometry enables viewers to relate the immaterial form of the model to material forms.

Conclusively, the principal development within parallel projection as it shifted from "isometrical perspective" to axonometric projection rests in the expressive freedom of an axially rotated view, while nevertheless maintaining scaled proportionality. Axonometry compels the free manipulation of the object in space so as to position the representation. Thus axial orientation is firmly tied to communicating perceptual aspects of the object as desired by the producer of the image. In so doing, axonometric projection aligns itself in a phenomenal way with conic perspective. Meyer and Meyer noted that in certain axial rotations, particularly trimetric projections, the difference between a perspective and axonometric projection is negligible, as the rotation of the axis system allows one to very closely approximate a point of view similar to conic perspective (Figure 3.5, left, comparisons with representations of a cross and an entablature).

A MATERIAL SENSE: THE OBJECT OF FARISH'S ISOMETRIC MODEL

Both models and pictures can be characterized by fidelity, or the degree to which they correspond to their referent. According to James Gibson (1904–79), a philosopher in the field of visual perception, faithful models "are defined as a physical object whose various surfaces have the same dimensions as the corresponding surfaces of the original object, and hence are geometrically congruent with them, but which is made of a different substance than the original," whereas a "faithful picture is a delimited physical surface processed in such a way that it reflects (or transmits) a sheaf of light-rays to a given point which is the same as would be the sheaf of rays from the original to that point."[37] In Gibson's articulation, one can argue that parallel projection drawings operate more as models than pictures, as parallel projection has greater fidelity to the object than to the visual perception of the object. As well, given Gibson's observation that "the most obvious kind of non-fidelity of a model is any distortion of a shape," one can argue that isometric projection—with its axial coordinates isometrically fixed and objectively depicting an object's material, spatial reality—has greater fidelity to the object than axonometric projection.[38]

The consequence of allowing unique, axonometric "points of view" is that the undifferentiated reality of the material object, as seen through the objectifying uniformity of the isometric drawing, is lessened. Today, of course, Farish's isometric projection is a specific case of axonometric projection, yet unlike all other angles of rotation, the isometric maintains a spatially uniform representation. Farish's isometric is parallel to the measured, spatial world of the material object. Isometry alone is tuned to objectively portray the material condition and, in doing so, operates as an iconic model by foundationally establishing a proportional image relating mimetically to the referent object. The isometrically construed model fulfills this end more clearly than any other form of pictorial representation since the act of positing an axonometric point of view introduces a sense of vision to the inherently nonoptical realm of parallel projection. Thus, in terms of pictorial representation and modeling, one must conclude that the isometric axis is tied to a conception of space tightly correspondent to the three-dimensional, physical object, and accordingly, the isometric drawing is an iconic model with a strong material sensibility despite being an immaterial, two-dimensional representation.

FARISH'S DEVICES AND DRAWINGS: MODELS FOR ENVISIONING IMMATERIAL AND MATERIAL REALMS

Models of material artifacts, and material culture as a discipline, share a foundational focus on objects as representational expressions. Farish presents us with multiple methods of modeling, principal among them a material (kit-of-parts) device used to illustrate forces and physical phenomena and a system for drawing developed to best model the spatial attributes of the material world. Farish used three-dimensional modeling devices to visualize, or analogically model, kinematic principles, but Farish's drawings also model the very methodology of modeling. His two-dimensional isometrical representations call attention to the protean nature of iconic models. By the same token that his isometry renders a spatial and scaled representation parallel to a material referent, his isometrical drawing technique offers a model for envisioning the critical material realities of the physical, material world.

While Farish's three-dimensional, composite modeling device was meant to reveal and model analogues of impermanent kinematic phenomena, it is through the demonstration device that Farish materializes the concept. Not only is the modeled phenomenon ephemeral but each configuration of the device that creates the model is ephemeral as well. The analogic model reminds us that models are not uniformly tangible or mimetic objects. Of course, Farish does engage in a visually mimetic form of iconic modeling. His fixed, two-dimensional isometrical images iconically modeled his protean modeling devices. Critically, isometry renders a uniquely object-oriented, spatial, and scaled representation parallel to its material referent, unlike other axonometric angles that are adjusted to a viewer. The isometric offers a mimetic model of the autonomous material reality of a physical object through drawing.

In sum, Farish's different modalities confirm that modeling is always a form of representation that requires interpretive choices. One must limit the communication of some information to heighten the communication of what is deemed essential. Farish's work testifies to the ways that drawing can model concepts and objects by distorting perspective, rather than practicing a kind of faithful realism. New forms of modeling arise when people must seek out new forms of representation to illustrate what is critical. Farish reminds us that shifts in the means of visioning and envisioning shift *what* is communicated by altering *how* it is communicated.

NOTES

1 William Farish, "On Isometrical Perspective," *Transactions of the Cambridge Philosophical Society* 1 (1822): 1–19.

2 Salahuddin Choudhury, "On Models," *Design Methods and Theories* 13, no. 2 (1979): 90–94. Choudhury notes that within an iconic model's visual structure, "Any quality other than these relations may be excluded without any effect. Transformation of this structure is possible as long as this visual relation is retained" (91).

3 Choudhury. Choudhury notes that within an analogic model's logical structure, "Any quality other than these relations may be excluded without any effect. Transformation of this structure is possible as long as this logical relationship is retained" (91).

4 *Oxford English Dictionary*, s.v. "natural philosophy, n." and "natural philosopher, n."

5 Kevin C. Knox and Richard Noakes, eds., *From Newton to Hawking: A History of Cambridge University's Lucasian Professors of Mathematics* (Cambridge: Cambridge University Press, 2003), 69–134; Stephen D. Snobelen and Larry Stewart, "Making Newton Easy: William Whiston in Cambridge and London," in Knox and Noakes, 135–70.

6 Gerard L'Estrange Turner, *Nineteenth-Century Scientific Instruments* (London: Sotheby, 1983), 69–70. William Whiston, Newton's protégé and immediate successor as the Lucasian chair, initiated the use of demonstration apparatus in the Newtonian physics course at Cambridge. After his dismissal in 1710, Whiston spread the didactic methodology, giving lectures illustrated with demonstration models in London.

7 Turner, 69.

8 Willem Jacob 's Gravesande, *Mathematical Elements of Natural Philosophy Confirmed by Experiments; or, An Introduction to Sir Isaac Newton's Philosophy* (1720, Latin), Engl. transl. J. T. Desaguliers (London: J. Senex and W. Taylor, 1720). Referenced here: 1747, 6th ed., vol. 1, "An Oration Concerning Evidence," xxxvi. Note: "Spoken at Leyden the eighth of February, in the Year 1724."

9 's Gravesande, xlvii–iii.

10 André-Marie Ampère, *Essai sur la philosophie des sciences, une exposition analytique d'une classification naturelle de toutes les connaissances humaines*, vol. 1 (1838), 51–52. Translated and quoted in Eugene S. Ferguson, *Kinematics of Mechanism* (Washington, D.C.: Smithsonian Institution, 1962), 209.

11 Ferguson, 209.

12 Robert Willis, *Principles of Mechanism* (London: J. W. Parker, 1841), vii. Willis recorded the first notable shift in approach within Jacob Leupold's *Theatrum Machinarum* (1724), which contained "the first attempt to consider the parts

of machinery separated from their work, and referred to the modifications of motion."

13 Willis, vii. These mechanical and geometrical parts are today known as mechanics and statics. Willis translates from a memoir and lauds Leonhard Euler, who in 1775 made "the first clear statement on the true principles upon which the science of Kinematics must be based" (viii).

14 Turner, *Nineteenth-Century Scientific Instruments,* 69.

15 See note 8; 's Gravesande, *Mathematical Elements,* mentioned within the preface to 1st ed., ix. Also Turner, *Nineteenth-Century Scientific Instruments,* 69.

16 Turner, *Nineteenth-Century Scientific Instruments,* 73. See also Florence Grant, "Reading, Writing, Drawing, and Making in the 18th Century Instrument Trade," *Science Museum Group Journal* 1 (2014), https://doi.org/10.15180/140103.

17 's Gravesande, *Mathematical Elements,* Plate 27.

18 T. J. N. Hilken, *Engineering at Cambridge University, 1783–1965* (London: Cambridge University Press, 1967), 51.

19 Sir Robert Stawell Ball, "Details of the Willis Apparatus Used in Illustrating the Foregoing Lectures," in *Experimental Mechanics: A Course of Lectures Delivered at the Royal College of Science for Ireland* (London: Macmillan, 1888), 345–53.

20 Farish obituary in *Memoirs of the Royal Astronomical Society,* vol. X (London: J. Weale 1838), 366–68.

21 Hilary Bryon, "Building a Modern Vision: Auguste Choisy's Graphic Constructs," PhD diss., University of Pennsylvania, 2005. Chapter 6, "English Isometrical Perspective," reviews the developments of isometry. See also Bryon, "Revolutions in Space: Parallel Projections in the Early Modern Era," *arq: Architectural Research Quarterly* 12, no. 3–4 (2008): 337–46.

22 Farish, "On Isometrical Perspective," 2.

23 Bryon, "Building a Modern Vision," 76.

24 Farish, "On Isometrical Perspective," 5–6.

25 Bryon, "Building a Modern Vision," 72.

26 Farish, "On Isometrical Perspective," 4.

27 Claude Bragdon, *The Frozen Fountain: Being Essays on Architecture and the Art of Design in Space* (New York: Knopf, 1932), 61.

28 See Joseph Jopling, *The Practice of Isometrical Perspective* (1833); Thomas Sopwith, *A Treatise on Isometrical Perspective* (1834); and Thomas Bradley, *Practical Geometry, Linear Perspective, and Projection* (1834).

29 For an overview of the development of axonometry in Germany, see Bryon, "Revolutions in Space."

30 Meyer and Meyer identify Weisbach's brief scientific article titled "Die monodimetrische und anisometrische Projectionsmethode (Perspective)," translated as "Method of Monodimetric and Anisometric Projections (Perspective)." Today these correlate to dimetric and trimetric forms of axonometric projection. See

M. H. Meyer and C. Th. Meyer, *Lehrbuch der axonometrischen Projectionslehre* (Leipzig: H. Haessel, 1855).

31 Meyer and Meyer, 13. First published as *Lehrbuch der Axonometrie,* vol. I (1852), vol. II (1853), vol. III (1855).

32 Bryon, "Building a Modern Vision," 92.

33 Meyer and Meyer, *Lehrbuch der axonometrischen Projectionslehre,* 13.

34 Meyer and Meyer, 8. They observed that perspective is an inappropriate appellation.

35 Bryon, "Building a Modern Vision," 94.

36 Meyer and Meyer, *Lehrbuch der axonometrischen Projectionslehre,* 96.

37 James J. Gibson, "A Theory of Pictorial Perception," in *Reasons for Realism* (Ann Arbor: University of Michigan Press, 1982), 249.

38 Gibson, 250.

4

"The Instructed Eye": What Eighteenth- and Nineteenth-Century Drawing Books Tell Us about Vision and How We See

CHRISTOPHER J. LUKASIK

For more than four hundred years, vision was understood as a passive, natural, and largely involuntary process in which an image was simply transmitted through the eyes to the brain. Over the past quarter century, however, cognitive neuroscience has revealed vision to be a complex process of active, latent (stored), and multiple interacting sensory representations distributed widely across several distinct regions of the brain. "Vision," as the neurobiologist Margaret Livingstone succinctly puts it, "is information processing not image transmission."[1] The eye's conversion of light into neural impulses—known as the retinal image—is only the first stage of visual information processing, and without additional information supplied by the brain, it remains open to countless possible interpretations. As a result, vision requires two additional stages of information processing in which the visual cortex analyzes those incoming "visual stimuli according to primitive features, such as vertical and horizontal elements, angles, and curves," before dispatching them to other parts of the brain by means of a massively parallel network where they interact with our "vast personal knowledge of self and the world."[2] Our prior knowledge of the world, whether acquired or inherited, not only shapes vision at the most fundamental level (as in pattern and facial recognition); it literally determines the things we see. As Francis Crick and Christof Koch explain, "although the main function of the visual system is to perceive objects and events in the world around us, the information available to our eyes is not sufficient by itself to provide the brain with its unique interpretation of the visual world. The brain must use past experiences (either its own or that of our distant ancestors embedded in our genes) to help interpret the information coming to our eyes."[3] What we see is thus actively created by us, "formed by an integration of immediate multi-sensory information, prior experience, and cultural learning."[4] To a significant degree, our brain sees the present it experiences through the past it knows.

Contemporary research from neuroscience may seem far removed from the antiquated pages of eighteenth- and nineteenth-century drawing books that are the subject of this essay, yet these texts, I want to suggest, can tell us much about how we see, because they contain detailed information about how we have been instructed to see. If "the essence of vision" is construction, as cognitive psychologist Donald D. Hoffman contends that it is, then drawing instruction books afford us an opportunity to consider the codes, practices, models, and discourses by which we have been instructed to constitute the visible and invisible, the legible and illegible, and recognize them as such.[5] Moreover, drawing books, like vision itself, not only entail an active construction of the visual; they often do so in terms that explicitly address the multiple sensory modalities (seeing and touching) and cognitive functions (imagination, memory, attention/visual selection, and observation) that we now know to be central to human vision. "By presenting a comprehensive program for the learning of drawing," drawing books supply users with a standard artistic vocabulary for representing the world and thus provide them with a schema, grammar, or way of seeing the world consistent with and informed by the visual order of their particular historical moment.[6] For this reason, most scholarship on drawing instruction books agrees that these texts teach readers not only *how* and *what* to draw but also *how* to see. Drawing books, in short, model vision.

While scholars like Barbara Stafford have already looked to the art of the eighteenth and nineteenth centuries for evidence of the "ecological character of human cognition" and how "images lay down the tracks that affectively activate our eyes and mind," this essay examines how drawing books from the same period model vision.[7] As individual artifacts, these drawing books offer discrete models for how to represent the world visually by providing their users with written instructions to follow and optical examples to copy manually. Yet, when considered collectively as a material social practice over time, they also document the historically variable conditions that delimit and structure the visibility of that world. Drawing books from this period, I argue, participate in a more general reconfiguration of the human sensorium of vision, one in which the primacy of the mind to eighteenth-century drawing instruction yields to greater emphasis on the eye and hand, on one hand, or the mind, on the other, depending on their nineteenth-century commercial or fine art application. This recalibration of the mind, eye, and hand within drawing instruction of this period, as I will discuss in the following pages, was accompanied by three other significant shifts as drawing instruction broadened its audience during the first half of the nineteenth century: (1) the separation of the iconic and indexical functions of drawing into commercial and fine art applications, (2) the reorientation of drawing practice within the more popular and commercially oriented books from the imagination of the individual mind to the production of an accurate transcription of the physical world through standardized graphic elements, and (3) an increasing emphasis on conceptualizing drawing in terms of a natural or universal language. These four theoretical shifts in drawing instruction, as I explore at the end of this essay, contributed to changes in drawing practice, particularly to how students were taught to draw and see faces. My remarks on these transformations in drawing instruction are based on a survey of seventy-five drawing books published in England and America between the years of 1750 and 1860, with more than half appearing in England and none of the American publications appearing until the start of the nineteenth century. The majority of these texts are drawing instruction manuals—which discuss drawing either exclusively or alongside other traditional arts, such as engraving, etching, or painting—with the remainder being copy books, which are manuals comprising almost entirely optical images, or large compendium-type books, which include, among many other subjects, individual chapters on drawing instruction within their encyclopedic pages.

The hundred years from the middle of the eighteenth century to the middle of the nineteenth century, which form the historical parameters of this essay's discussion of drawing books, also mark a significant moment in the history of vision and visuality itself, because this was the time when modern culture became visual. It was during these years, as a number of prominent art historians have argued, that the new scopic regimes, artistic practices, and types of observers constitutive of a modern visual culture first emerged.[8] As Peter De Bolla explains with respect to the British context, some time in the middle of the eighteenth century, "something recognizable as precisely a *culture* based on the visual, on modalities of visualization, the production and consumption of visual matter (representations, maps, diagrams), as well as any number of mechanical objects intended for use in some form of looking, observing, surveying, spying, and so forth, all requiring and producing various modes of address, attention, or forms of understanding—that all this came together in the ways that theretofore had not resulted in a coherence or a coalescence such that it became possible to identify something called visual *culture*."[9] This new culture of visuality, De Bolla contends, was accompanied by new modes of looking across the eighteenth century as artistic production shifted from patronage to commercial networks of exchange. Similarly, the beginning of the nineteenth century is when Jonathan Crary claims that "a new kind of observer took shape in Europe radically different from the type of observer dominant in the seventeenth and eighteenth centuries."[10] Unlike "the incorporeal relations of the camera obscura" of centuries past, a new valuation of visual experience grounded in the physiological status of the observer and of vision emerges between 1810 and 1840 in which "the human body . . . becomes the active producer of optical experience."[11] The historical arc of the drawing books under discussion in this essay not only coincides with the period during which modern culture became visual; their theories and practices speak directly to the kinds of transformations in vision, visual culture, and observation that these scholars discuss. The practices of looking articulated within eighteenth- and nineteenth-century drawing books promise to provide additional historical specificity to these accounts of the emergence of a new modern visual culture and the formation of a new modern observer. In addition, the enormous popularity of amateur drawing during the first half of the nineteenth century uniquely positions these books to supplement our knowledge of visuality as a cultural practice of everyday life. Finally, because these books disclose the material practices and historical conditions under which things become visible, they offer us valuable documents of how vision was modeled during the period in which modern culture became visual.

The rise of drawing instruction books and amateur drawing across the eighteenth and nineteenth centuries, as a number of scholars have remarked, signals an important cultural phenomenon in Europe and North America. As Diane Strazde observes, "never before had so much stress been placed on the practical knowledge of drawing, nor had so many instruction books been published to serve these ends."[12] Formerly the provenance of the early modern male courtier, drawing as a social practice expanded "in the eighteenth century when, for the first time, its claims to be considered a polite and useful art were taken seriously" by both men and women of the upper and middle classes.[13] The introduction of drawing into British civilian education at the close of seventeenth century and the significance of landscape sketching to topography, theories of the picturesque, and the practice of touring over the course of the next hundred years, as Ann Bermingham has noted, further dispersed drawing and brought it into the bourgeois culture of civic and moral self-improvement.[14] No longer considered an activity reserved for the wandering eye of the aristocratic *virtuoso* or the delicate hand of the aspiring, but nonetheless leisured, artist, drawing's general utility and seemingly endless applications were increasingly described as a necessity by the middle of the eighteenth century. "There is scarce any Art or Profession which receives not some Assistance from Drawing" the *New and Compleat Drawing-Book* proclaimed in 1751. "Without her help, no Designs or Models can be well executed; to her, the Mathematician, Architect, and Navigator is indebted; no station of Life is exempted from the Practice of it."[15] Such broad assertions of drawing's utility were common within the prefatory material of drawing instruction books during the second half of the eighteenth century. They, along with claims for drawing's civic and national value, undoubtedly facilitated the spread of amateur drawing instruction over the next fifty to seventy-five years.

By the first quarter of the nineteenth century, however, amateur drawing was

ANY ONE WHO CAN LEARN TO WRITE CAN LEARN TO DRAW and, as writing is not taught to those only who are destined to become authors, but as forming an essential part of general education, so is drawing equally important to others besides professional artists. To write —to draw a form or figure that shall be recognised as the representative of a letter or word, is one thing; and to be able to design, draw, or write such forms, upon principles of grace and accuracy—to understand the Art of writing—is another. Thus it is also with Drawing, another mode of expressing ourselves, not less useful or necessary than that by letters

FIGURE 4.1. John Gadsby Chapman, *The American Drawing-Book; a Manual for the Amateur* (New York: J. S. Redfield, 1847). Courtesy of the Library Company of Philadelphia.

no longer just a polite and useful art; it was a burgeoning commercial market. The popularization and commercialization of drawing instruction grew to be so widespread during the nineteenth century that no fewer than 145 drawing manuals would be published between 1820 and 1860 in the United States alone.[16] With American public schools including drawing instruction in their curricula during the late 1830s, drawing was "second only to writing as an essential means of communicating practical information" for many mid-nineteenth-century Americans.[17] Drawing instruction, its nineteenth-century advocates argued, was useful for virtually anyone, not because it aided in the acquisition of an individual, artistic style, nor because it signified a superior social status, but because drawing developed the skills of observation and precision necessary for accurate penmanship, engraving, mapmaking, surveying, industrial design, and countless other jobs involving manual labor (Figure 4.1).

The popularization of amateur drawing that took place from the middle of the eighteenth century to the middle of the nineteenth century was accompanied by shifts in drawing instruction that speak to how students were taught to perceive and draw objects and thus reflect how drawing books modeled vision at the time. During these years, emerging notions of commercial and fine art practice separated the iconic and indexical functions of drawing that had formerly coexisted within early modern drawing instruction. Early modern drawing books had conceived of drawing as both an *iconic* sign (signifying a visible object in the world through resemblance and contributing to the production of scientific visual knowledge through the creation of truthful examples and models) and as an *indexical* sign (signifying the individual style or "hand" of the artist and supporting humanist concerns with individuality).[18] This dual function of early modern drawing—indexically pointing back to the genteel draftsman, on one hand, and iconically pointing forward to the empirical object of their attention, on the other (as in Oliviero Gatti's 1619 *Three Model Heads*)—was pervasive in the seventeenth-century *virtuosi,* "who looked at pictures as either signs of learning or signs of personal status."[19] Where these two functions had coincided in the drawings of the early modern courtier, over the course of the eighteenth century, they are reorganized—and frequently separated from each other—often in response to or in conversation with distinctions between fine art and mechanical reproduction, mental and manual labor, on one hand, and drawing's status as a leisure activity for the refined elite or a useful professional skill for the middle and working classes, on the other.

As publishers downwardly distributed drawing more deeply within Anglo-American society over the eighteenth and nineteenth centuries, this separation of the iconic and indexical functions of early modern drawing and their respective distribution to commercial and fine art practice was accompanied by a recalibration of the mind, eye, and hand within drawing instruction. The indexical function came to be associated with the mind of the artist (in distinction to and in control of her hand), fine art practice, high culture, individual style, and refinement. In contrast, the iconic function came to be associated with the eye and hand (in distinction to and often independent of the mind), professional or craft practice, low culture, utility, and mass production/consumption. As late as the second half of the eighteenth century, despite appeals to its broad utility, drawing remained largely a product of the genteel mind. The mind's ability to judge and remember, for instance, was indispensable to Gérard de Lairesse's drawing instruction. Whatever success the student has in drawing, he argued, "will be more owing to his Genius than his Hand."[20] "The Hand must be employed in the practice," de Lairesse says, but "the Mind in Judging."[21] While the draftsman should be "an exact copier of Nature," this practice was clearly understood as a mental activity more than a manual one (a distinction undoubtedly important to preventing drawing from devolving into a merely mechanical exercise). Indeed, de Lairesse advises pupils to take "a close view of your original, and in your Mind, divide it into several Parts" before setting pencil to paper.[22] Carington Bowles, whose drawing instruction books went through multiple editions during the last quarter of the eighteenth century, continued to circulate de Lairesse's instructions, suggesting that "the hand will be more ready to execute those objects the mind has so clearly conceived."[23]

Yet, the mind's role in drawing was critical not just for learning how to draw but for learning how to see. In his 1772 *The Draughtsman's Assistant,* for instance, Bowles insists that drawing instruction was valuable even for "gentlemen" who don't draw at all, because it teaches them how to look at and judge the work of others. As a result, these gentlemen become "connoisseurs, in the polite arts of painting, engraving, &c."[24] Bowles's emphasis on the mind shielded drawing instruction from charges that it was little more than mindless copying. Similar to the kind of eighteenth-century looking activity described by Peter De Bolla as part of the era's "regime of the picture," late eighteenth-century drawing books such as Bowles's taught the student how to look at the same time that it instructed them in how to draw.[25]

By the start of the nineteenth century, however, amateur drawing was no longer simply for the discriminating mind of the connoisseur; it was for everyone. As publishers marketed drawing books to a larger and less genteel audience in the nineteenth century, the more corporeal modalities of drawing—eye and hand—took precedence over the mind in drawing instruction intended for commercial rather than fine art practice. Rembrandt Peale's 1835 *Graphics,* for instance, which had "the largest circulation of any American drawing book before the Civil War," promised its students that drawing would correct the eye's perceptions and instill manual dexterity.[26] Similarly, John Gadsby Chapman's 1847 *The American Drawing Book* emphasized that "the precision and facility of the hand and the certainty of touch" that the draftsman would acquire through drawing "will enable him to wield the crayon or the brush, the graver or the modelling tool, the chisel or the hammer, with a command that will amply repay the labor of his present efforts to become familiar with it."[27] Writing at a time when English manufactures were flooding into

American markets, Chapman attributed this situation to the lack of artistic education in the American workforce. For this reason, he proposed teaching drawing not only to mechanics and their assistants but to lower-class women—those "dependent females who are compelled to toil"—and even prostitutes in an effort to improve the design and craftsmanship of their manual labor.[28] In Chapman's pedagogy, the "facility of the hand" takes priority over the "conceiving mind" that was so central to eighteenth-century drawing instruction.[29] This reconfiguration of the human sensorium in drawing instruction—one made largely within the context of the commercial as opposed to fine art applications of amateur drawing during the first half of the nineteenth century—confirms the emergence and instrumentalization of a new kind of modern observer grounded in corporeal subjectivity, as Jonathan Crary suggests, but drawing's more corporeal modalities of eye and hand were emphasized to mute rather than amplify the presence of subjective vision. Drawing's value, for pedagogues such as Chapman, for example, consists in how it "practices the eye to observe; and the hand to record."[30] For Chapman, the mind retreats into the eye as the optical supplants mental or imaginative activity in drawing. In fact, the opening series of drawing lessons in *The American Drawing Book*—which do not even contain images, only mere lines to be traced and copied—are designed to develop the hand's precision, distinction, and clarity so that drawing's iconic and communicative function in accurately representing objects could be mastered.

In contrast, other mid-nineteenth-century drawing instruction books, particularly those that understand drawing as a preparatory component of fine art instruction, decouple the indexical from the iconic function of drawing by rejecting precisely the same elements that were being emphasized in more popular drawing books marketed to working-class readers. Books such as Fielding Lucas's lavish 1827 *Progressive Drawing Book*, J. D. Harding's 1834 *Elementary Art; or, The Use of the Chalk and Lead Pencil*, and Frank Howard's 1841 *The Sketcher's Manual*, which conceived of drawing instruction as a fine art, typically in preparation for either watercolor or painting, criticize the emphasis on the eye and hand in drawing pedagogy as well as the use of the copying method in the development of the amateur draftsman. No matter how "acute their sight may be, or ready their hand for facsimile imitation," Harding's *Elementary Art* declares, these draftsmen "will never be able, of themselves, to select and truthfully depict the beauties of Nature."[31] Harding asserts that the study of art is ultimately "an appeal to the mind, and not, as commonly supposed, to the eye only."[32] He claims that students need to "cultivate the faculties of observation, comparison, and reflection, so that whilst the mind is thus acquiring more distinct ideas of forms and their relations, it be prepared to perceive and appreciate the higher beauties with which Nature has surrounded us."[33] Too often, students "can do no more than follow the mechanical progress of the work with his eye, without being able to follow the Artist in his mind."[34] In contrast to Peale and Chapman, the end of drawing for Harding was to produce "as near a likeness to Nature . . . not so much by laborious copy addressed to the eye only, as by reviving in the mind those ideas which are awakened by a contemplation of Nature."[35]

Frank Howard's *The Sketcher's Manual* similarly complains that "so many works upon the Art of Drawing . . . simply tell pupils to copy plate A or plate B."[36] In these books, Howard laments that "everything is left to the eye" and that the student "has no knowledge of what are the essential qualities of a picture."[37] Consequently, his sketches lack "the charm of Pictorial Effect," by which Howard means "that quality which distinguishes a Picture from a diagram or map."[38] Howard's explicit contrast of "a picture" with "a diagram or map"—both iconic signs in Peirce's sense—demonstrates how fine art drawing instruction was distinguishing itself from commercial art by moving away from its iconic function. As Howard clarifies in a later chapter, in a drawing with pictorial effect, the objects are not simply seen but, in strikingly Lockean terms, leave an "impression on the mind."[39] Recalling the familiar terms of eighteenth-century drawing books, Howard's fine art drawing speaks to the mind more than the eye, and it foregrounds drawing's indexical function by disclosing the artist's knowledge of the "essential qualities of a picture." In this way, the separation of the iconic from the indexical in drawing instruction during this period maintained long-standing divisions between the genteel and common by separating the allegedly higher-order functions of mental activity from the presumably lower-order ones of the manual/optical activity and assigning them to the artistic individual, on one hand, and the mass laborer, on the other.

As fine art drawing books retreated into older notions of drawing as primarily directed by the mind and as having primarily an indexical function, more popular drawing books reoriented drawing as a practice away from the visualization of the imagination of an individual mind to the production of an accurate optical transcription of the physical world through standardized graphic elements. This reorientation was undoubtedly related to the shifting definitions of drawing as primarily a mental, ocular, or manual activity discussed earlier. As the examples of Peale and Chapman suggest, by the middle of the nineteenth century, the mind's role in popular drawing practice was gradually diminished by increasing attention to an almost mechanical training of the eye and hand. Consequently, the mental activity of drawing, if it was discussed at all, came to be articulated more in optical than imaginative terms (except in the domain of explicitly fine art instruction). Where mid-eighteenth-century drawing instruction books, such as Robert Sayer's popular 1755 *The Compleat Drawing Book*, define the object of drawing as whatever "observation can discover or Imagination conceive," by the end of the century (with the exception of the picturesque book for landscape sketching), drawing instruction books would refer to the imagination far less, if at all.[40] Consider, for example, de Lairesse's 1751 definition of drawing as "a representation, by lines and shade, of any Object in Nature or Art; the Imitation of another Draught, or the expressing the Likeness of something conceived of our Mind" with the less imaginative definition offered by Carington Bowles in his 1785 *The Artist Assistant in Drawing*, in which drawing is now "the art of representing by outlines and shadows, the various productions of nature and art, and of enlarging and contracting objects in the most exact proportions."[41] Here the work of drawing is understood less in terms of imaginative genius than in terms of producing scale models.

By the time Peale's *Graphics* was published in 1835, the mental activity of drawing was almost entirely conflated with the optical. To be sure, the mind remained important for fine art drawing instruction, but for drawing books with more commercial applications—and they reached a far larger audience with their many editions—the eye had replaced the imagination. For authors like Peale and Chapman, the mind was no longer understood as the fountainhead for conceptualizing or imagining a drawing that would then take shape through the eye and hand on the page. Instead, it largely served as an information processor between eye and hand. "The art of drawing," Peale explains, "requires a knowledge of the forms and proportions of objects, and the practice of marking them on a plane surface, as they might be marked on a glass held between the eye and the objects."[42] In drawing's ideal state, the indexical presence of the individual's mind and the medium of drawing should simply disappear, just as the glass does in Peale's metaphor, so that the spectator experiences nothing but the accurate transcription of visible objects. For Peale, drawing demands the immediacy of the image, which, in turn, depends on how accurately the eye and hand can reproduce that physical world without registering the materiality of the medium or the presence of the artist's hand.

Moreover, the accuracy of a drawing is not the result of prior cognitive and perceptual abilities already in the mind of the individual draftsman. Instead, those abilities are acquired through the optical and manual proficiency that drawing instills. Drawing, Peale argues, provides "the advantages of correct perception and accurate discrimination to the educated eye, and the power of exact definition and precise demonstration alone to the experienced hand."[43] Whereas at the beginning of this period, the individual mind of the draftsman was understood as the origin of the drawing—whether in terms of rendering a visible form from the physical world outside it or visualizing an imaginative form from within it—now the "instructed eye" was the effect of drawing.[44] Just as eighteenth-century British visual culture generated looking practices—which Peter De Bolla associates with the scopic regimes of the picture (based on recognition) and of the eye (based on identification)—these more commercially oriented drawing manuals taught many readers how to perceive objects in the world according to basic graphic elements before instructing them to draw those objects according to those same graphic elements. Nineteenth-century popular drawing instruction thus utilized the kind of recognition involved in the prior regime of the picture but repurposed it toward basic graphic elements. Students were taught to see objects in terms of the basic graphic elements from which they were composed. At the same time, and in the interest of communicative efficiency, it also stripped that activity of looking as a means of locating individual identity. Drawing was not so much to be seen as read, but what it made legible, at least for most amateurs, was neither the high cultural narratives of the regime of the picture (which persisted in fine art drawing instruction) nor the subjective identifications of the regime of the eye but the standardized graphic elements of the visual.

This reorientation of drawing practice away from originating in the individual mind to becoming instead the product of an "instructed eye" and hand not only contributes to the deindividualization of drawing practice and diminishes the role of the imagination in it but facilitates the reconceptualization of drawing as primarily communicative in its operation. "Drawing," as Peale describes it, "is little more than writing the forms of objects."[45] Peale's *Graphics* models vision as a process in which the visible is constituted in terms of standardized and reproducible units. These basic, standardized visual elements—such as "perpendicular, horizontal, diagonal, and circular lines"—correspond with many of the primitive visual elements that the brain uses to detect patterns and process incoming visual stimuli into coherent images.[46] Similarly, Peale's students learn to perceive objects according to these primitive parts in order to depict them accurately in their own drawings. Through this process, students learn how to move the hand "in all directions" and train "the eye to judge whether those directions are correct."[47] As a result, Peale's drawing instruction also instills a homogenization of drawing practice and, in doing so, standardizes what counts as an "accurate" perception. To represent the physical world through standardized visible elements first necessitates constituting the visible according to those elements and recognizing them in the world. Students must learn how to model the world according to those basic linear elements of what D. B. Dowd refers to as the "glyphic mode" rather than simply using their own eyes.[48] In this sense, the corporeal subjectivity of an individual observer—that retinal image seen from the uninstructed eyes—did not serve as the basis for mimesis in amateur drawing instruction in its commercial as opposed to fine art context. Instead, students learned how to see the physical world through standardized visible elements to arrive at a version of the mimetic maximized for communicative efficiency and reproducibility. It is worth noting how this standardization of visual representation in drawing practice developed independently of the advent of photographically based optical media that would later mechanize this same process.

Drawing instruction's emphasis on the production of an accurate optical transcription of the physical world by training the eye to recognize standardized graphic elements, as mentioned, privileged drawing's communicative function moving forward. In fact, from the last quarter of the eighteenth century forward, and with increasing frequency, drawing books, no matter their fine art and commercial contexts, often conceptualized drawing in terms of a language. The conception of drawing as a universal language, however, was paradoxically predicated on asserting drawing's media specificity in terms of its sensory mode—visual as opposed to verbal—and its signification system, iconic as opposed to symbolic. While the conceptualization of drawing as a language is evident at the beginning of this period, if not earlier—Benjamin Franklin, for instance, refers to drawing as a "universal language" in 1749—it is important to note that this articulation is ultimately grounded in understanding drawing as primarily an optical medium that nonetheless speaks to the eye.[49] Drawing acts as a kind of universal language for Franklin because it is "understood by all Nations. A man may often express his Ideas to his own Countrymen, more clearly with a Lead pencil, or a bit of Chalk, than with his Tongue."[50] For Franklin, drawing's capacity to communicate ideas clearly through iconic signs seen by the eye makes it superior to the conventional symbolic signs of language heard by the ear. Moreover, this remains the case even when people speak the same language. In his 1764 *The Principles of Drawing*, de Lairesse advances a similar point when he describes drawing as a "universal language, understood by all men."[51] Yet, for de Lairesse, drawing's advantage as a universal language rests not merely in how it signifies in an apparently unmediated way but also in its ability to communicate ideas that words simply cannot express. Drawing "represents to our view, the forms of innumerable objects we have no other ways of beholding; and helps us to the knowledge of many of the works in nature or art, which any other method of describing would be insufficient to give an idea of."[52] Drawing, in other words, conveys an informational density superior to language. Carington Bowles registers both of these points in his 1785 *Bowles's Artist Assistant in Drawing* when he describes how drawing, "the silent, but most expressive language of nature, which speaks to the eye, is understood by all nations, and conveys an idea where even words themselves would prove deficient."[53] For these authors, drawing is similar to a universal language in terms of its ability to communicate ideas clearly, but it is unlike a language in that its sensory mode and signification system are not liable to the limitations of language (chiefly, the arbitrariness of its signs and the insufficiency of their descriptive power in the face of the immediacy and informational density of a drawn image).

The notion that drawing could serve as a universal language recalls the universal language schemes of seventeenth-century linguists such as John Wilkins, Comenius, and John Willis. Whether it was Comenius's reliance on pictures, Wilkins's hopes for a universal language, or Willis's shorthand system, all the linguists of this period agreed "that ideas can be visually represented, contained in the contours of lines on a page, or evoked by the distribution of agreed-upon marks."[54] Yet, what is remarkable is that the articulation of drawing as a universal language remained consistent throughout the eighteenth and nineteenth centuries. The drawing manuals of the late eighteenth and early nineteenth centuries—such as those authored by William Marshall Craig, Samuel Prout, David Cox, and J. D. Harding—continued Wilkins's attempt to devise a universal language well into the first half of the nineteenth century.[55] Craig's attempt, for example, to devise a system of natural signs sought a resemblance between the "graphic sign and the referent, as opposed to the abstract and arbitrary relationship."[56] "What Craig envisioned," Ann Bermingham explains, "was a natural sign—a visual sign that would maintain a highly legible and transparent correspondence to the thing it represented, and which for this reason would be impervious to manipulation, change, and misunderstanding because it would be locked in a mirror-like relationship to its referent."[57]

By the nineteenth century, however, the conceptualization of drawing as a universal language shifted from differentiating its sensory mode and signification system from those of language (in that pictures were thought to communicate more information, more clearly than words) to now conflating them with each other. For authors like Peale and Chapman, drawing and writing were now defined as analogous tasks: "Writing is nothing else than drawing the forms of letters. Drawing is little more than writing the forms of objects."[58] Objects, in other words, were the alphabet of this universal language, and one only had to learn to recognize/see their component graphic elements to read them and reproduce them clearly in pictures for others. The desire to understand drawing as a kind of universal language—routinely expressed in so many nineteenth-century drawing instruction books—contributes to understanding the iconic and symbolic operations of drawing as equivalent processes and thus allows for the arbitrariness of the latter's semiotic meaning to be concealed through the resemblance of the former's sensory mode and, as a result, to mute the role of the meaning-making mind, in Peircean terms, in establishing the relationship between a symbol and its signification.

Despite addressing drawing largely in abstract and general terms, these theoretical shifts in drawing instruction from this period—the evolving recalibration of the mind, eye, and hand within drawing instruction; the separation of the iconic and indexical functions of drawing into commercial and fine art applications; the reorientation of drawing practice within the more popular and commercially oriented books from the imagination of the individual mind to the production of an accurate transcription of the physical world through standardized graphic elements; and an increasing emphasis on conceptualizing drawing in terms of a natural or universal language—also contributed to changes in material drawing practice, particularly in drawing the face. From the last quarter of the eighteenth century forward, there is an extension of that conceptualization of drawing as a universal language to the representation of the face in ways that conflate the codes for rendering a face legible with the face itself. During this period, the ways in which drawing books model human difference and understand the social functions of those visible human differences shift from depicting individual faces as variations of a universal, albeit idealized and Eurocentric, human figure to inscribing the supposedly visible and permanent traces of individual or national character, or the signs of ethnic or racial difference, on the face.

Prior to the last quarter of the eighteenth century, in books such as Lens's 1751 *A New and Compleat Drawing-Book,* de Lairesse's 1752 *The Principles of Drawing,* and Sayer's 1755 *The Compleat Drawing Book,* the first and most important step in drawing the human face was always to register its correct form and proportions. In the "Fourth Lesson" of de Lairesse's *The Principles of Drawing,* for example, the pupil is advised to sketch the head first so that he can "secure the proportions in the most exact manner."[59] In drawing the face, de Lairesse recommends beginning with a basic shape, such as an oval, and then dividing the shape proportionately so as to conform to the prevailing aesthetic demands of beauty (Figure 4.2). The "perpendicular in the oval being divided into four equal parts," he notes, "makes the whole head to be four noses in height."[60] Once the student learns how to draw the "cross of the oval" line for the face, he can then draw a variety of faces in any number of positions. The instruction to draw from a basic shape (such as an oval) to the particular face also supports the period's more general embrace of imaginative drawing practice noted earlier, because it enables the student to draw an infinite number of faces, including those "out of your own invention or fancy."[61] Even copy books composed entirely of images, such as Sayer's *The Compleat Drawing Book,* reproduce the same pedagogy by providing students with engraved models (Figure 4.3) from which to draw generic heads before dividing the face into symmetrical parts. The optical model thus first trains students to recognize shared human form and proportion before registering individual variations within it. In these works, there is a standard, basic shape from which the draftsman proceeds in which the incalculable variety of human faces are understood not in terms of identity and difference so much as in resemblance and variation.

Although the eighteenth-century aesthetic norms of symmetry and proportion

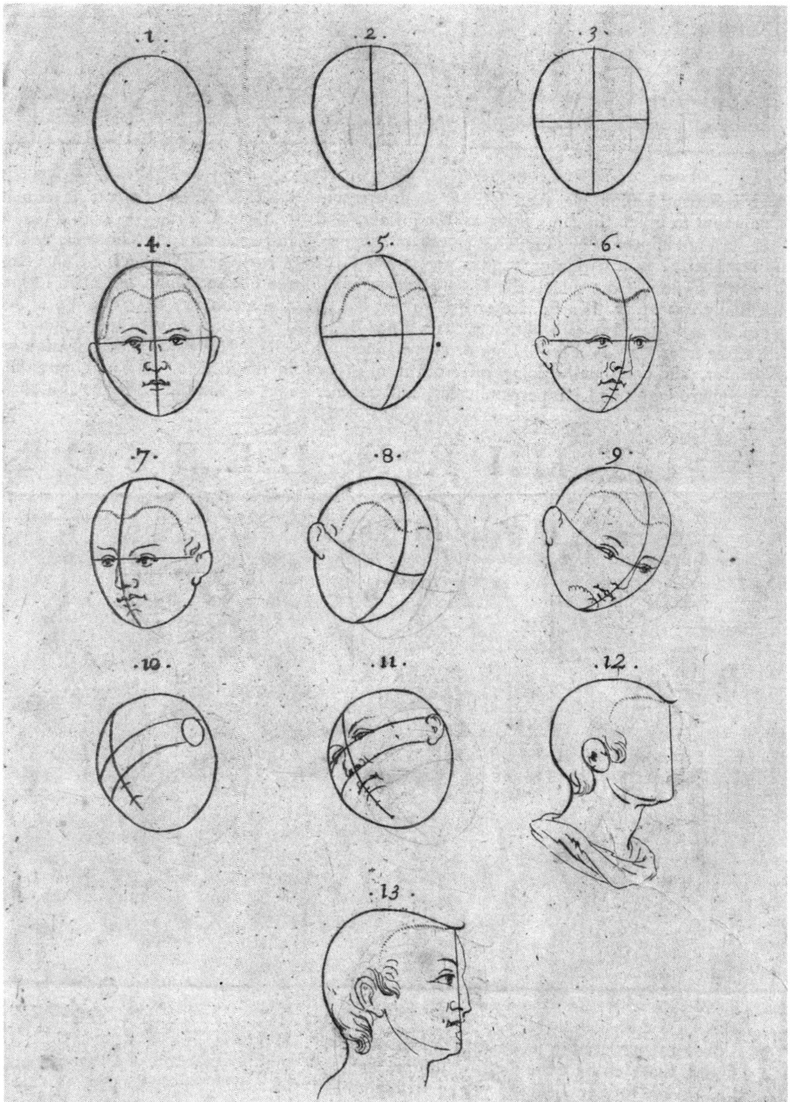

FIGURE 4.2. Gérard De Lairesse, *The Principles of Drawing,* 5th ed. (London: Thomas Bowles, 1752), 6. Courtesy of University of Michigan Clements Library.

for the human face would remain consistent features of drawing instruction well into the nineteenth century, the subordination of the individual variations of a face to those aesthetic demands would dissipate across the first half of the nineteenth century as drawing instruction became more concerned with registering the marks of individual or national character or the signs of ethnic or racial difference on the human face. This shift had much to do with the rise of Lavaterian physiognomy

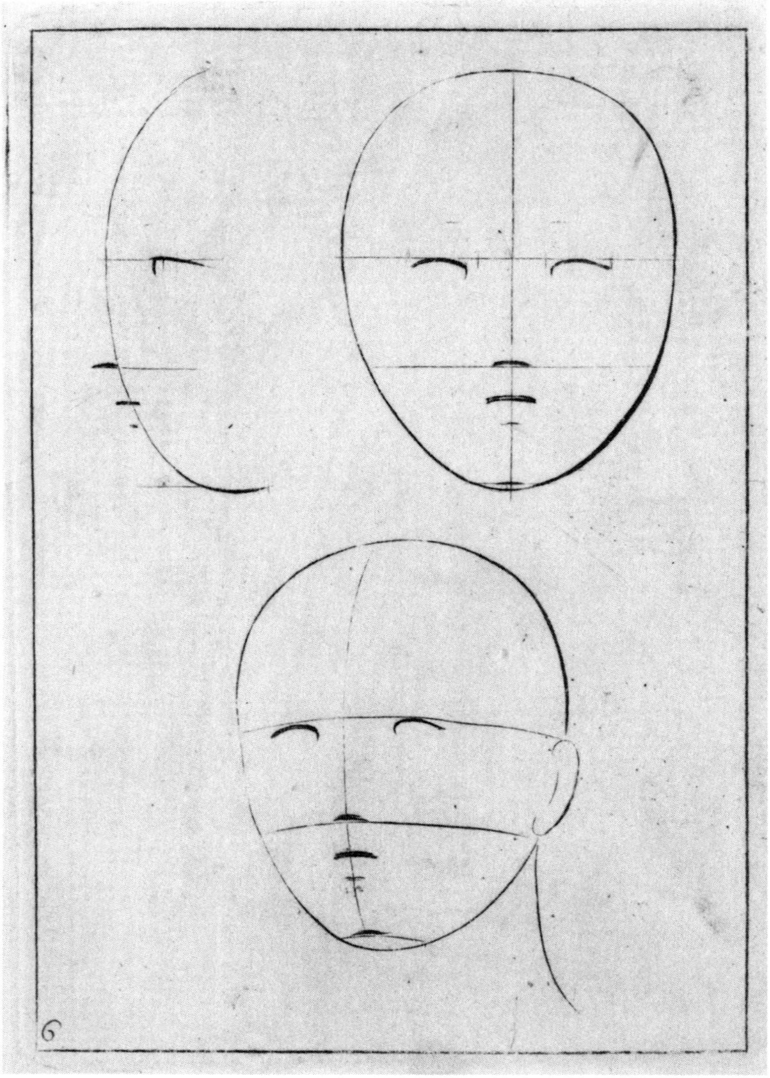

FIGURE 4.3. Plate 6 from *The Compleat Drawing Book* (London: Robert Sayer, 1755). Courtesy of the Library Company of Philadelphia.

at the end of the eighteenth century and its practice of discerning moral character from the particularities of the face. Unlike Giambattista Della Porta's *De Humana Physiognomia* (1586) and Charles Le Brun's *Conférénce sur l'Impression des Différents Caractères des Passions* (1668)—which were the leading eighteenth-century manuals for depicting a person's passions and her visible, but impermanent, facial expressions—Lavater's *Essays* (1789) depicted the correspondence between a person's permanent moral character and her unalterable facial features (such as the shape and length of the nose).

Drawing was at the center of Lavater's model of physiognomic observation, and the two practices mutually informed each other. "Drawing is the natural language of Physiognomy," Lavater asserted in the *Essays.* "It is . . . the only medium of fixing with certainty, of portraying, of rendering sensible an infinite number of signs . . . which it is impossible to describe in words."[62] For this reason, Lavater insisted that no one should be admitted to "the study of our science" unless he demonstrated "*talent* . . . in the practice of drawing."[63] Lavater was an avid draftsman himself, and his *Essays* closely "followed the established drawing-manual tradition of selective and sequential division or disembodiment of the figure," with particular attention to the component parts of the face.[64] Similar to drawing books, Lavater's *Essays* taught its readers how to perceive the face according to basic linear elements, literally tracing the signs of supposed moral character from it.

The tremendous popularity of Lavater's *Essays* throughout the nineteenth century—fifty-five editions had already been published by 1810—not only distributed this model of physiognomic observation widely but also transformed how faces were drawn by emphasizing particularity and traditionally low forms—such as drawn silhouettes and profile portraits—over idealized faces and oil portraiture as the preferred media and styles for representing the face.[65] In his lectures delivered before the Royal Academy, for example, Henry Fuseli rejected the suppression of particularity championed by his predecessor and rival Sir Joshua Reynolds by stating that physiognomy, "the companion of anatomy," was indispensable for drawing the body.[66] Similarly, more commercially oriented drawing books, such as Charles Hayter's *An Introduction to Perspective, Drawing, and Painting* (1815), Nathaniel Whittock's *Oxford Drawing Book* (1845), George Child's *A New Drawing Book of Figures* (1845–52), and Chapman's *American Drawing Book,* either specifically instructed students to study Lavater or reproduced his stylistic emphasis on outlines and clear, hard profiles.

By the middle of the nineteenth century, and particularly in America, however, the discourse of physiognomy was yielding to more modern conceptions of racial difference, and drawing books from this period would register that change. Physiognomy as a scientific discourse for understanding the meaning of the face was being challenged by more quantitative, if equally arbitrary, discourses, such as phrenology and the scientific racism of the American school of anthropology. In this context, the attention to proportion in drawing the human face was repurposed from conforming to the Eurocentric aesthetic ideals of beauty or disclosing permanent moral character to delineating the physical, measurable differences among racial types.[67] Louis Bail's 1859 *Bail's Drawing System,* for instance, explicitly recommends that "phrenology should always be studied in connection with physiognomy" when drawing the human face. Bail's drawing book contains engravings of "the heads of the different races of men."[68] In the two outlined heads, identified as "Model Caucasian" and "African Beauty," respectively, Bail visually inscribes the differences, which are then described verbally as perceptually self-evident (Figure 4.4). The head of the "Model Caucasian," Bail remarks, "is very nearly perfect; the intellectual region of the head fully developed,"[69] whereas "African Beauty" "may be considered a good specimen of the head of a native African; in common phrase, 'a smart, likely negro.'"[70]

In an effort to ground these differences into the phenotypical features of the face, Bail racializes physiognomic observation, noting how "the angle of the Caucasean [*sic*] face is very obtuse, in fact, it approaches a straight line. That of the African face, is nearly a right angle. The form of the head exhibits a contrast, if possible, yet more striking, and proves the value of the study of the skull, and also the beauty, symmetry, and dignity, its correct formation gives to the whole man."[71] Similar to the drawing instruction of Peale and Chapman, Bail's student is taught first to look for and recognize the basic graphic elements constitutive of the form—in this case, those signifiers of race—to be drawn. To draw a barrel, for example, authors such as Peale recommended that the student first see and recognize the circles and lines of which it consists. The student would, in other words, perceive the object by actively constructing it through her prior knowledge of those basic visual elements (such as shape and line) before then drawing it according to those same elements. Likewise, to see and draw the face, Bail suggests that the student must first recognize the racial angles and proportions of which it consists.

Like Bail's drawing system, J. G. Heck's drawing instruction in the 1851 *Iconographic Encyclopedia of Science, Literature, and Art* also demonstrates how the racial types generated by phrenology and scientific racists like Samuel Morton, Josiah Nott, and George Gliddon inform drawing practice and model vision. When it comes to drawing the human head, Heck's students are not told to begin with a universal shape, such as an oval, and then inscribe individual variations but rather to start with each race's distinct profile proportions. "In the Caucasian race," Heck notes, point "g stands back from [point] b about half the length of the nose; while in negroes it advances almost two thirds the length of the nose."[72] As the examples of de Lairesse and Sayer, on one hand, and Heck and Bail, on the other, demonstrate, the arc of learning how to draw the face during this period reflects more general changes in how the signs of moral character and race were deposited into and then rendered visible on the face before being read.

In this way, drawing instruction developed a model of vision consistent with and informed by a range of discourses (physiognomy, phrenology, and racial

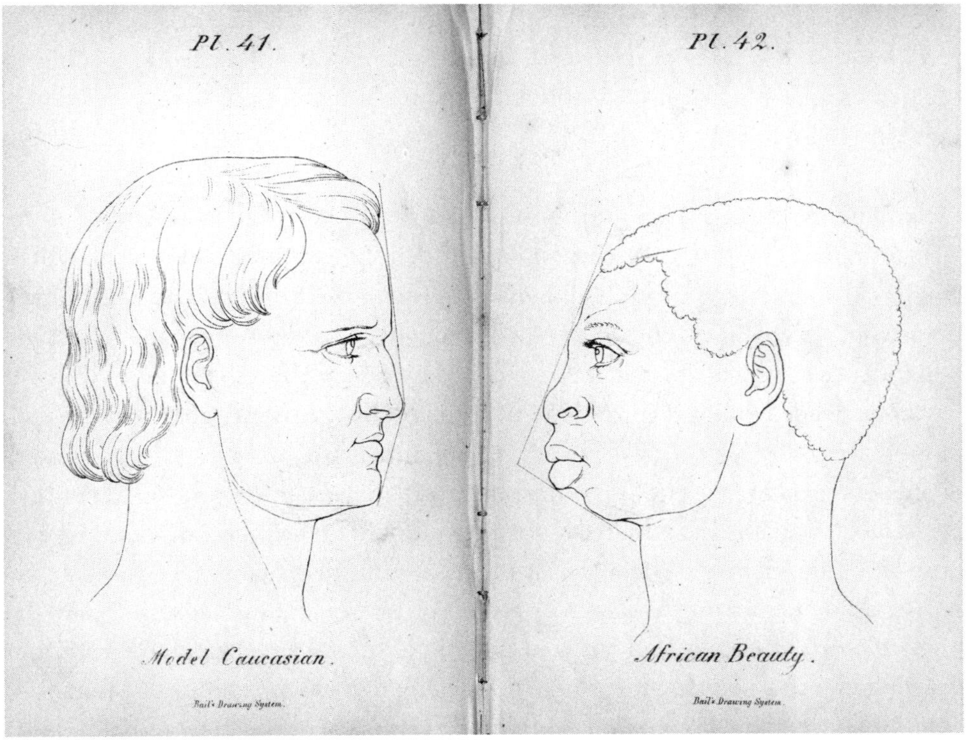

FIGURE 4.4. Plates 41 and 42 of Louis Bail, *Bail's Drawing System. The Human Head: A Correct Delineation of the Anatomy, Expressions, Features, Proportions and Positions of the Head and Face* (New Haven, Conn.: L. Bail, 1859). Courtesy of the Library Company of Philadelphia.

science among them) that sought to locate the phenotypical features of race before "looking" at anyone in particular even transpired. Students looked for and recognized race in the face first in Bail's system just as they had looked for and recognized the basic oval shape in the face during the previous century. Of course, Bail's system, like Lavater's mode of physiognomic observation, creates the very faces that it claims to objectively observe, draw, and interpret. As Francoise Delaporte explains, "the measure of the facial angle and the curve of a forehead are not signs, but they become so from the moment when character deposits—or rather, transposes—its qualities onto the face."[73] The drawing instruction of the mid-nineteenth century thus extends an earlier practice of looking—or what I have identified elsewhere as "physiognomic discernment"—in which the essential features of a person's individual and permanent character or collective identity were known in advance and assigned to the face before retroactively reading it there afterward.[74] Moreover, since this practice of looking was positioned as a model for drawing the face, it undoubtedly informed and legitimated a wide range of popular racist visual culture in which the faces of individuals were first read and depicted according to the perceived signifiers of race, ethnicity, and gender that were used to construct them in the first place.

I would like to close this essay around two passages that speak to a number of the transformations I discussed earlier while also identifying some developments still to come. The first returns us to Chapman's *The American Drawing Book.* In it, Chapman asserts that "however exalted be the thought . . . , it must be made to assume a shape by which it can be conveyed and understood beyond the mind in which it was conceived. Whether words, letters, or forms, be the means of expression employed, *they must be intelligible*; to make them intelligible, they must be accurately expressed, in a language not to be mistaken; and *that accuracy is no man's intuitive possession.*"[75] Chapman's remarks return to a number of the points discussed previously: the alignment of the iconic with the communicative function of drawing; the reorientation of drawing toward the production of an accurate transcription of the physical world through standardized graphic elements; the conceptualization of drawing in terms of a language and a privileging of that communicative function; and the assertion of drawing's sensory mode (visual as opposed to verbal) and its signification system (iconic as opposed to symbolic) in that conceptualization. Yet, what makes this passage worth noting is the way in which the eye must learn to see according to the model by which the visible will be produced. Chapman insists that no single person intuitively possesses this model of visibility. Much like our contemporary neuroscientific understandings of vision, much of what we come to see is the product of what we have learned to see. As Chapman's example suggests, drawing instruction books generate models by which visible things become legible. It is in this sense that drawing books model vision. To put it another way, the "seeable" paradoxically depends on the intelligibility of an image, not on its visibility.

This importance of legibility—whether it be basic standardized visual elements or the signifiers of racial types—to the visible during the period when modern culture became visual is also discussed in Francis Grose's 1788 *Rules for Drawing Caricaturas.* In it, Grose advises that "when a caricaturist wishes to delineate any face he may see in a place where it would be improper or impossible to draw it, he may commit it to his memory, by *parsing* it (as the school-boys term it) by naming the contour and different features of which it is constructed, as school-boys point out the different parts of speech in a *Latin sentence.*"[76] For Grose, as it was for many other drawing instruction book authors at the time, the technique by which one sees and then draws a face is analogous to the process of reading. The face is a sentence comprising distinct grammatical parts—the sweep of the forehead, the line of the nose, the shape of the lips—that signify discrete physiognomic meanings (intelligence, morality, sensuality) that, when read together, communicate social meanings as much as they convey a figurative resemblance. In this way, the visibility of the face has less to do with optical resemblance than with the historical conditions by which the face can be read and thus seen in the first place. The visible world is one that is ultimately legible. To be unseeable here, however, is not to be invisible; rather, it is to be illegible according to Grose's physiognomic system (even if that system was still based on Della Porta's soon to be antiquated physiognomics of animal resemblance).

Grose's remarks not only anticipate how later nineteenth-century drawing manuals will deploy similar models in which the component parts of the face are made to speak to differences in national character and race, as we saw in Heck and Bail; they also disclose its primary social context. The point of compartmentalizing the face into grammatical units, Grose explains, is for those moments when one is unable to draw it—such as on the street. In those locations where it is "improper or impossible to draw" faces, he explains, the spectator should translate the face into those grammatical parts, store the image in her mind, and draw it later. Of course, the assumption was that only the discerning few—as opposed to the undiscerning many—would be capable of such mental activity, but it demonstrates how such legibility also allows for a type of portability, surveillance, and control. The division of the image of the face into its component parts allows the spectator to categorize, record, and store that face in her memory wherever she may be and without detection. This model of drawing, in other words, allows for acquisition, a form of possession in the Sontagian sense, and it does so long before its technical perfection through the medium of photography or the technology of biometrics. As William Robson put it a decade after Grose in his 1799 drawing manual *Grammigraphia,* drawing "allows us to possess an image," which, in turn, allows us to view objects or persons "in safety."[77] "Drawing communicates a universal acquaintance with every visible object," and in doing so, Robson continues, "it gives a certain degree of possession."[78] Drawing under this system of visibility does not simply replicate the original; it captures it.

By way of conclusion, I would like to return to the question with which this essay began: what might the history of drawing books have to tell us about how we see? As I hope the preceding pages have demonstrated, eighteenth- and nineteenth-century drawing books tell us—long before neuroscientists reached the same conclusion—that we actively construct what we see and base those constructions on what we already know and how we have been instructed to see according to that knowledge. Situated at the origins of our modern visual culture, eighteenth- and nineteenth-century drawing books demonstrate that we may only see what is recognizable or legible; that those recognitions are acquired culturally and through material models; that the visible is delimited to those historical articulations of the legible; and finally, that the point of those recognitions, at least in the emergent mid-nineteenth-century context, may be to take possession of visibility in ways that reproduce and ensure the privileges and pleasures associated with the vantage point that is empowered to articulate its legibility in the first place.

On the basis of these points, the study of eighteenth- and nineteenth-century drawing books may also prompt us to move beyond the familiar division of visual culture and material culture studies along the axis of an opticality understood apart from its materialist foundation. The multiple sensory modalities and cognitive functions at the heart of this period's drawing instruction—with their intertwined, evolving emphasis on eye, hand, and mind—should alert us to the limitations of understanding images exclusively in terms of opticality or as a material artifact independent of the discourses, codes, and practices by which it becomes legible. Visuality and materiality need not be oppositional, and as Nick Yonan suggests, "it is possible to imagine visual culture and material culture as interrelated aspects of the same scholarly project."[79] Indeed, the example of eighteenth- and nineteenth-century drawing books reveal how the material practices for drawing instruction and the discourses for understanding what was drawn mutually informed each other. The drawing book's integration of immediate sensory information from a material object, with its optical models to copy manually, and its integration of prior cultural knowledge—through a variety of aesthetic and scientific discourses—not only mirrors the multiple stages of human vision; it may also provide a template for how the fields of material culture and visual culture studies might work together in addressing the materialist and constructivist aspects of an image.

NOTES

1 Margaret Livingston, *Vision and Art: The Biology of Seeing* (New York: Abrams, 2002), 53.

2 Robert L. Solso, *The Psychology of Art and the Evolution of the Conscious Brain* (Cambridge, Mass.: MIT Press, 2003), 6.

3 Quoted in Barbara Stafford, *Echo Objects: The Cognitive Work of Images* (Chicago: University of Chicago Press, 2007), 142.

4 Ann Marie Seward Barry, *Visual Intelligence: Perception, Image, and Manipulation in Visual Communication* (Albany: State University of New York Press, 1997), 15.

5 Donald D. Hoffman, *Visual Intelligence: How We Create What We See* (New York: W. W. Norton, 2000), 10.

6 Chia-Chuan Hsieh, "The Emergence and Impact of the 'Complete Drawing Book' in Mid-Eighteenth-Century England," *Journal for Eighteenth-Century Studies* 36, no. 3 (2013): 410.

7 Stafford, *Echo Objects,* 17 and 11.

8 See Jonathan Crary, *Techniques of the Observer* (Cambridge, Mass.: MIT Press, 1990); Ann Bermingham, *Learning to Draw: Studies in the Cultural History of a Polite and Useful Art* (New Haven, Conn.: Yale University Press, 2000); and Peter De Bolla, *The Education of the Eye* (Stanford, Calif.: Stanford University Press, 2003).

9 De Bolla, *Education of the Eye,* 4.

10 De Bolla, 6.

11 Crary, *Techniques of the Observer,* 16, 69.

12 Diana Strazdes, "The Amateur Artist and the Draughtsman in Early America," *Archives of American Art Journal* 19, no. 1 (1979): 15.

13 Bermingham, *Learning to Draw,* 77.

14 Bermingham characterizes the period from 1745 to 1825 as being dominated by landscape drawing, which she then divides into three kinds: a landscape of sense (associated with topography and factual transcription); a landscape of sensibility (associated with the picturesque mode and idealization); and a landscape of sensation (associated with the artist's own perceptual experience). Whether pursued in a utilitarian way or as a polite amusement, drawing served a variety of "individual, civic, and national purposes," leading Bermingham to conclude that the discourse surrounding drawing in the eighteenth and nineteenth centuries is "highly complex" (78).

15 Bernard Lens, *For the Curious Young Gentlemen and Ladies That Study and Practise the . . . art of Drawing, . . . a New and Compleat drawing-book; . . . Being the close study, . . . of the late Mr. Lens, . . . in sixty-two copper-plates, engraved by himself . . . To which is prefixed, an introduction to drawing; . . . Translated from the French of Monsieur Gerrard de Lairesse, and improved with extracts from C. A. Du Fresnoy* (London: B. Dickinson, 1751), 1, Gale-Cengage Eighteenth-Century Collections Online.

16 Peter Marzio, "The Art Crusade: An Analysis of American Drawing Manuals, 1820–1860," in *Smithsonian Studies in History and Technology,* no. 34 (Washington, D.C.: Smithsonian Institution Press, 1976), 1.

17 Elliot Davis, "Training the Eye and the Hand: Drawing Books in Nineteenth-Century America," PhD diss., Columbia University, 1992, 10.

18 See Bermingham, *Learning to Draw,* 45. My sense of the distinction between iconic and indexical signs follows that of C. S. Peirce in his seminal essay "Icon, Index, Symbol." For Peirce, the icon signifies by virtue of its resemblance to the object itself. Thus mimesis and imitation are iconic. Furthermore, he claims that "the icon has no dynamical connection to the object it represents; it simply happens

that its qualities resemble those of that object, and excite analogous sensations in the mind for which it is a likeness. But it really stands unconnected with them" (114). In contrast, an index "is a sign, or representation, which refers to its object not so much because of any similarity or analogy with it, nor because it is associated with general characters which that object happens to possess, as because it is in dynamical (including spatial) connection both with the individual object, on the one hand, and with the senses of memory of the person for whom it serves as a sign, on the other" (107). "The index," Peirce explains, "is physically connected with its object; they make an organic pair, but the interpreting mind has nothing to do with this connection, except remarking it, after it is established" (114). In the context of drawing, a drawing may be said to be iconic to the degree to which it resembles its referent; it may be said to be indexical to the degree to which it refers back to the style or "hand" of the artist.

19 Bermingham, 47.

20 Lens, *For the Curious Young Gentlemen,* 2.

21 Lens, 2.

22 Lens, 4.

23 Carington Bowles, *The Draughtsman's Assistant; or, drawing made easy. Wherein the principles of the art are laid down in a familiar manner, in ten lessons . . .* (London: T. Kitchin, [1772]), 5, Gale-Cengage Eighteenth-Century Collections Online.

24 Bowles, *Draughtsman's Assistant,* 5.

25 De Bolla claims that during the emergence of a culture of visuality in eighteenth-century Britain, there were two poles of debate over visual culture: (1) a commitment to an egalitarian politics of taste and (2) a familiar investment in the superiority of the knowing viewer. From these two positions, De Bolla identifies two scopic regimes—the "regime of the picture" and the "regime of the eye," with two different "kinds of looking activity." *Education of the Eye,* 16. In the "regime of the picture," looking is based on the recognition of a name, style, genre, or painter within an artistic tradition, which then enables the viewer to identify with a particular subject position, which De Bolla identifies as the connoisseur. In the more egalitarian "regime of the eye," the process of recognition is suspended while identification takes place. *Education of the Eye,* 18. One learns to look only by looking itself.

26 Marzio, "Art Crusade," 20.

27 John Gadsby Chapman, *The American Drawing-Book; a Manual for the Amateur . . .* (New York: J. S. Redfield, 1847), 36.

28 Chapman, 8.

29 Chapman, 11.

30 Chapman, 4.

31 James Duffield Harding, *Elementary Art; or, The Use of the Chalk and Lead Pencil,* 4th ed. (London: Day and Son, 1854), 2.

32 Harding, 7.
33 Harding, 2.
34 Harding, 8.
35 Harding, 13.
36 Frank Howard, *The Sketcher's Manual, or*... (London: Darton and Clark, 1841), advertisement.
37 Howard, advertisement.
38 Howard, advertisement.
39 Howard, 3.
40 *The Compleat Drawing Book* (London: Robert Sayer, 1755), 4.
41 Lens, *For the Curious Young Gentlemen,* 1 ("a representation"), and Carington Bowles, *Bowles's Artist Assistant in Drawing*..., 7th ed. (London: Printed for the author, [1785?]), 9, Gale-Cengage Eighteenth-Century Collections Online ("the art of representing").
42 Rembrandt Peale, *Graphics: A Manual of Drawing and Writing for the Use of Schools and Families* (New York: J. P. Peaslee, 1835), 6.
43 Peale, 6.
44 Peale, 6.
45 Peale, 6.
46 Peale, 7.
47 Peale, 6.
48 D. B. Dowd, *Stick Figures: Drawing as a Human Practice* (St. Louis, Mo.: Spartan Holiday, 2018), 15.
49 Benjamin Franklin, *Proposals Relating to the Education of a Youth in Pennsylvania,* in *Writings,* ed. J. A. Leo Lemay (Washington, D.C.: Library of Congress, 1987), 329.
50 Franklin, 329.
51 Gérard de Lairesse, *The Principles of Drawing*..., 6th ed. (London: John Bowles, 1764), 1.
52 Lairesse, 1.
53 Bowles, *Bowles's Artist Assistant in Drawing,* 9.
54 Murray Cohen, *Sensible Worlds: Linguistic Practice in England, 1640–1785* (Baltimore: Johns Hopkins University Press, 1977), 14.
55 Bermingham, *Learning to Draw,* 110.
56 William Marshall Craig, *Essay on the Study of Nature in Drawing* (London: W. Bulmer, 1793), 108.
57 Bermingham, *Learning to Draw,* 109.
58 Peale, *Graphics,* 6.
59 Gérard de Lairesse, *The Principles of Drawing*..., 5th ed. (London: Thomas Bowles, 1752), 6, Gale-Cengage Eighteenth-Century Collections Online.
60 Lairesse, 6.
61 Lairesse, 6.

62 Johann Caspar Lavater, *Essays on Physiognomy . . .*, trans. Henry Hunter, 3 vols. (London: John Murray et al., 1789–98), 1:122.

63 Lavater, 1:390.

64 Ross Woodrow, "Lavater and the Drawing Manual," in *Physiognomy in Profile: Lavater's Impact on European Culture* (Newark: University of Delaware Press, 2005), 74.

65 On the impressive publication history of Lavater's *Essays*, see John Graham, "Lavater in England," *Journal of the History of Ideas* 22, no. 4 (1961): 562.

66 Henry Fuseli, *Lectures on Painting: Delivered at the Royal Academy March 1801 / by Henry Fuseli, P.P.: with Additional Observations and Notes*, 2 vols. (London: H. Colburn and R. Bentley, 1830), 2:18.

67 See David Bindman, *Ape to Apollo: Aesthetics and the Idea of Race* (Ithaca, N.Y.: Cornell University Press, 2002), 225.

68 Louis Bail, *Bail's Drawing System . . .* (New Haven, Conn.: L. Bail, 1859), 19.

69 Bail, 19.

70 Bail, 19.

71 Bail, 19.

72 J. G. Heck, *Iconographic Encyclopedia of Science, Literature, and Art*, 4 vols. (New York: Rudolph Garrigue, 1851), 4:151.

73 François Delaporte, *Anatomy of the Passions*, trans. Susan Emanuel (Stanford, Calif.: Stanford University Press, 2008), 39.

74 Christopher Lukasik, *Discerning Characters: The Culture of Appearance in Early America* (Philadelphia: University of Pennsylvania Press, 2011), 35.

75 Chapman, *American Drawing Book*, 24, emphasis original.

76 Francis Grose, *Rules for Drawing Caricaturas: with An Essay on Comic Painting* (London: A. Grant, 1788), 14, Gale-Cengage Eighteenth-Century Collections Online.

77 William Robson, *Grammigraphia; or, The Grammar of Drawing* (London: Printed for the author, 1799), 23.

78 Robson, 23.

79 Michael Yonan, "Toward a Fusion of Art History and Material Culture Studies," *West 86th: A Journal of Decorative Arts, Design History, and Material Culture* 18, no. 2 (2011): 239.

5

Algorithmic Audition: Modeling Musical Perception

MARTIN SCHERZINGER

> Ideology gets modeled in software.
> —Alex Galloway, *The Interface Effect*

EXECUTION WITHOUT EXPLANATION

What follows is a sketch of a recent generation of music application software that attempts to model human perceptual faculties. Software applications are knowledge-bearing devices with a particular physiognomy. They proffer new forms of attention. This chapter will argue that auditory perception, modeled by market-driven technologies, is subsequently reshaped in the image of that technology. The chapter will suggest that the kind of modeling at work in these new technologies undoes some conventional expectations about models—that they are either prospective/predictive or mimetic/imitative. In other words, the music software under consideration is a model of human perception that in fact reshapes, in ideological terms, the human sensorium it purports merely to represent. The chapter will show how the perceptual feel of music—down to its beats and its grooves—is in fact entangled in processes of technical, economic, legal, and industrial mediation.

Notwithstanding the persistent debate on the subject of music's ephemerality and ineffability, most musicologists, theorists, and ethnomusicologists today construe music as material culture.[1] This is particularly true for music when it is considered as an interaction between technical equipment and a technique of sensing, and even more so when it is considered in light of the tangible traces that compose its digital representation. By deploying various algorithmic techniques—frequently described in anthropocentric terms like "machine listening," "machine cognition," and "machine learning"—the software under consideration segments audio signals into computer-measurable units that ostensibly chart tangible musical features.[2] In addition to extracting traditional features from songs—such as notes, harmonies, timbres, pitches, dynamics, vocal syllables, and metric beats—this software deploys algorithms oriented toward cultural information about songs, gleaned from extensive databases. Examples include natural language processing and collaborative filtering techniques. These algorithms harvest cultural and contextual tags about songs, extracted from online text associated with songs, including music reviews, social media, blogs, comment boxes on streaming platforms, and the like. The combination of content-based analytics and natural language processing techniques (deriving information from datasets) is today mobilized for music services in ways that transcend their most immediately apparent applications. In other words, these services do not only generate playlists, connect users, track rights management, facilitate browsing, curate musical tastes, and assist composing, as one might expect, but increasingly model fundamental musical experiences and sensations by simulating human perception itself. For example, the founders of Echonest—a "music intelligence company" (founded in 2005 and acquired by Spotify in 2011)—boast that their "dynamic" music data set (covering millions of songs and billions of music data points) deploys "principles of psychoacoustics, music perception, and adaptive learning to model both the physical and cognitive processes of human listening."[3] In short, with access to billions of statistically formatted data points, music's new generation of application software has effectively devised algorithmic models for simulating the workings of the listening ear.

Using recent "groove tracking" and "beat tracking" technologies as a central referent, this essay offers a theoretical reflection on the epistemological stakes of the recent turn toward an algorithmic control model of sensory experience. In an era characterized by a plethora of biotechnological innovations—a research tradition that deploys living sensory systems to develop technological applications—we have witnessed the ascendance of various traditional scientific disciplines (cognitive psychology, sociology, neuroscience, etc.) for machine learning. The essay describes and assesses the various components implicit in the perceptual models that are assembled by groove and beat trackers to forge a new amalgam between the human body and the algorithmic machine. The argument is twofold. First, the essay shows that the algorithmic body projected by the software into its decision-making procedures is conceived in atemporal rationalist terms—attendant to measurable statistical correlations and regularities extracted from large sets of data. Second, the essay demonstrates that these software products gently channel the extreme diversity of human sensory behavior that characterize global cultural practices into a generalized Euro-industrial counterpart. Using music from Africa as an argumentative punctum, the paper gestures toward the perceptual limits of these products. Although colonial musical standards have long held sway over the general representation of musical practices across the globe, the automation of such representations in embodied, interactive software applications shifts the scope of their authority.[4]

Although it is tethered to infrastructures of transnational capital, biotechnical research is in fact nurtured under specifically localized socioeconomic conditions, with the result that engineers and developers become firmly embedded in equally localized social forms of human sensory engagement. A specific set of social and cultural values—codified in the form of sensory information, or data points that are the input and output of their own process—is thereby layered and cemented into lines of code. The adoption of networked wireless devices—connected by microprocessors and sensors that register and record embodiments of real-time sensory feedback in milliseconds—becomes a fertile arena for anticipating human perception within a feedback loop. A technical form of anticipatory attending—forecasting human perception at a microtemporal level—thereby composes (in both senses) variations in sense perception. Far from probing the plasticity of perception, or the variety of its cultural inflection, one may speak here of the modification of human perception by streaming sensors on networked subjects. In short, the *modeling* of human perception in this arena is also a *construction* of human perception, which—in the context of automated media—has the capacity to *modify* perception. What follows is a theoretical reflection on the kind of computational knowledge we find in cognitive models for machine listening, with a particular interest in how these models comport actual sensory engagement and musical listening within a technoperceptual feedback loop. In other words, once the software is meaningfully incorporated into practice—from compositional procedures for automated music to music applications for everyday activities; from therapeutic aids for degenerative diseases to standards for litigating copyright claims—users also shape and rearticulate it in a kind of recursive loop between a calculating human and a calculating machine, with each adjusting to the other.[5]

When it comes to machine listening, the age of digitization has brought about a qualitative mutation in research methods across a variety of disciplines. On one hand, it is important to recognize that historical precedents for the technical modeling of human perception are a constitutive component for understanding artificial intelligence (AI) research today. For example, Fred Lerdahl and Ray Jackendoff's *Generative Theory of Tonal Music* (1983) was an influential text that intersected aspects of neuroscience, music theory, linguistics, and cognitive psychology. The theory laid out a "generative" model for human perception—including cognitively hardwired preference rules for human meter formation—that was simultaneously symbolized as a kind of proto-computational set of algorithmic instructions.[6] It may therefore be misleading to ascribe a paradigmatic break between the so-called digital age and its apparently antecedent predigital one. On the other hand, historical links between current models for the quantification of sensory reflexes and their historical precursors are not always what they seem. The combination of unprecedented computing power, vast troves of digitized training data (harvested within a globally networked architecture), and novel algorithmic training protocols significantly have marked a break with the past. Where AI was once construed in terms of generative, symbolic, and human-readable systems that required an explicit and declarative elicitation of rules and procedures, deep neural architectures could learn complex representations, and derive abstractions from their training sets, in a versatile statistical manner. The older form of AI—dominant until the 1980s (and exemplified by Lerdahl and Jackendoff's generative model)—was sometimes referred to as an "expert" system (with a "brute force" design), in which a computer was instructed to follow a lengthy set of instructions and then assigned to draw conclusions by applying various combinations of those instructions.[7] When it came to modeling human cognition, reasoning, or sensing, however, the older rule-governed algorithmic approach was generally met with limited success. Broadly speaking, no set of computational instructions was exhaustive or flexible enough to simulate the unique character of human thought and experience. The current flourishing of research into AI—from computer vision to speech recognition by way of robotic touch, taste, smell, and, of course, musical listening—was the result of a decisive shift *away* from either the formal modeling of the human sensory-motor system or the generative structure of the human linguistic system and toward large-scale statistical data processing.[8] By the late 1980s, the idea that computers had an internal kind of logic set apart from human-derived expert disciplines became axiomatic. In the memorable words of the electrical engineer Frederick Jelinek, "We thought it was wrong to ask a machine to emulate people. After all, if a machine has to move, it does it with wheels—not by walking. If a machine has to fly, it does so as an airplane does—not by flapping its wings. Rather than exhaustively studying how people listen to and understand speech, we wanted to find the natural way for the machine to do it."[9] Instead of working with syntax, rules, and grammar, Jelinek proposed training computers to recognize repeated patterns and derive statistical probability.

The shift away from formal modeling and toward statistical tools held equal sway in the case of computational music processing. While speech processing is a research field with a long tradition, music processing is a strikingly young discipline. Arguably, the statistical turn in AI was in fact foundational for the inauguration of an autonomous discipline, known as music information retrieval (MIR), with little historical precedent. The International Society for Music Information Retrieval, established in 2000, was thus coterminous with the widespread dissemination of internet-enabled digital computers and the emergence of large banks of digital data. Computer-based research into music had hitherto been conducted mostly within a symbolic framework of representation, including systems ranging from music notation to the musical instrument digital interface (MIDI) descriptors developed in the early 1980s. The exponential increase in digitized audio material stored in massive online data banks—along with a dramatic increase in computing speed, memory, bandwidth for information processing, and so on—facilitated statistical paradigms for the automated processing of waveform-based audio signals, unshackled by the traditional symbolic representations of music in code.

Far from simply modeling the sensorium by way of rule-governed symbolic systems, then, human capacities could increasingly be derived from an array of computational techniques that permitted machines to seek out patterns autonomously in ever-expanding data sets. This disciplinary turn away from rule-governed simulations of cognition and perception, and toward statistical data processing as such, became the defining hallmark of AI in the early twenty-first century. As a result, a new research paradigm reframed the very idea of machine cognition—including seeing, hearing, touching, and smelling—as a fundamentally *computational endeavor,* radically untethered from the very human faculties it sought to emulate. In the field of machine listening, for example, the declarative tree structures of the 1980s—expert systems grounded in the generative structures of speech and meter perception—gave way to statistical analysis and data processing, or the prediction of probabilities and estimates grounded principally in correlations detected within large databases. While delinked from any *explanation* about the phenomena it modeled, the strength of the statistical turn in AI lay in its power to *materialize* those phenomena. In other words, the machinic semblances of human cognition produced by seemingly unfettered (and even unsupervised) data processing systems surpassed those produced by systematic representations designed to model it. Again, the shift from representational to statistical modeling had less to do with the turn to digitization itself than with a turn toward an old method newly innervated by an exponentially expanding set of materially available data. In short, for statistically grounded AI, a cognitive model was less *represented* than it was *realized.* The model of human perception in these software applications takes the form of *task accomplishment* rather than *understanding* and thereby directs perceptual plasticity and cultural variety into a statistically mandated order. Put another way, in the world of AI, algorithmic actions do not *explain* sensory perception as much as they *execute* it.

ALGORITHMIC EPISTEMOLOGY (FROM KANT TO HEGEL-IN-REVERSE)

What kind of knowledge is sedimented into AI software? What is its method, scope, and validity? Let us take a closer look at some recent music application software that attempts to model human perception. These technological tools were designed to reduce the rigidity of various standardized templates that characterized a previous generation of digital tools for music. MIDI, for example, which had become the dominant representation scheme for music in code in the 1980s and 1990s, deployed standardized sound banks coordinated by "event messages" and "clock signals" to specify notation, pitch, and rhythm in musical time.[10] For example, a symbolic representation of a message using 1s and 0s, such as <11111010>, could indicate that the machine start playing a given sequence—a message that would then be followed by timing clocks that could trigger a message for stopping that sequence. MIDI was an on–off/up–down keyboard-like conception of music that depended on a host of technical standardizations. In 1983, the MIDI Manufacturing Association provided the standardizations that described a protocol, a digital interface, and connectors that enabled a variety of instruments, computers, and devices to connect and communicate with one another. To accomplish this, further standardizations, such as the default MTS (MIDI Tuning Standard, or conventional Western equal temperament) in 1992, ensured the elimination of variation in note mapping and thereby created the conditions for the possibility of interoperability on a global scale. However, as the MIDI protocol gained in popularity in music studios around the world, audiophiles, academics, and musicians simultaneously lamented its standardized templates. From critical academic commentary in the pages of *Computer Music Journal* in the 1980s by writers and programmers like Richard Moore and Nicola Bernardini to the process-inflected arguments by Jon Drummond and others at the Institut du Recherche et Coordination Acoustique/Musique in Paris, the meteoric rise of MIDI was met with an outpouring of popular and academic writing that castigated its limits. Broadly speaking, this critique identified three limitations in MIDI: first, the zoetrope-like mosaic approach to musical parameters induced by MIDI; second, the aleatoric microtiming problem of MIDI, involving signal jitter, temporal smearing, and other forms of nonhuman indeterminacy; and third, the unidirectionality of the communicative flow of most MIDI connectors.[11]

A new generation of software applications aspired to release creative musical technologies from the discrete triggering signals of MIDI and to enjoin compositional practice to a more temporal continuous signal representation.[12] Against this keyboard-centric conception of music, which construed music on the model of discrete quanta, the recent growth in music informatics now paved the way for a more fluid (or "intelligent") engagement with music—a user experience that was perceptually attuned to the continuity of the sound signal. This turn was grounded in the science of music information retrieval, a discipline that draws on a hybrid set of analytical methods—from sociological sentiment analysis, opinion mining, and social network analysis to cognitive psychology, musical language processing, machine learning, and neuroscience—to retrieve data from music.

With the emergence of massive databases for music's audio features (such as the Million Song Dataset, introduced in 2011 by Echonest and LabROSA at Columbia University, and the Music Genome Project, an ongoing project begun in 2000 by Will Glaser and Tim Westergren), MIR has played a considerable role in designing the software that is shaping new kinds of musical practice today.[13] Of particular interest to commercial and noncommercial stakeholders are systems for categorizing music, including recommendation software for streaming services (such as Spotify, Amazon Music, Apple Music, YouTube, and so on); song recognition software (such as Shazam); automatic transcription software (such as Sibelius and Finale for MIDI inputs); music applications attuned to weather, mood, activity, geolocation, and a host of social events; and digital tools for manipulating and creating music that can adapt to listeners' gestures and activities in real time. Generally speaking, then, current research in MIR attempts to map various types of human–computer interaction by designing technical interfaces for a seamless human encounter with music. The central challenge in this arena involves the computational extraction of subjective human sensory experience, or how listeners hear features like pitch, harmony, and rhythm in music, and entrainment patterns, or how listeners synchronize to felt grooves, affective states, and metric beats in music, by way of signal processing techniques. In short, the research attempts to model the human auditory system in code.

At this point, it may be useful to draw a distinction between technologies that *model* human capacities and those attempting to *enhance* them. In the world of computing, for example, Joseph C. R. Licklider, director of the Defense Advanced Research Projects Agency's Information Projects Techniques Office in the mid-1960s, differentiated between two types of human–machine interaction. On one hand, Licklider advanced the idea of mechanically extended man, and on the other, he cautioned against automated AI: "Mechanical extension has given way to the replacement of men, to automation, and the men who remain are there more to help than to be helped."[14] Licklider's distinction hinges on the difference between technological prosthetics, augmentations, and enhancements, on one hand, and AI designed to replicate human cognition, on the other. Jonathan Grudin, a computer scientist at Microsoft Research, for example, describes human–computer interaction (HCI) as the antithesis of AI: the aim of HCI was to "get away from studying human behavior and consider the computer as a tool for solving certain classes of problems."[15] At stake for these writers is a kind of historical struggle between augmentation technologies and AI and the attendant elimination of human capacity by the computing systems they build on the model of human cognition. The music software under consideration in this chapter falls squarely in the category of AI.

In basic philosophical terms, one may map such a distinction onto a notion of "Kantian" technologies and "Hegelian" ones. Immanuel Kant argued that human knowledge has its own internal conditions, which objects had to satisfy to become objects of knowledge. Given the formal conditioning ground of all knowledge (which included space, time, causality, and so on), Kant posited the possibility of a world beyond cognition—a *noumenal* world of things in themselves. Georg Wilhelm Friedrich Hegel's philosophical project, in contrast, eliminated the subject–object dualism implicit in Kant's construal and outlined instead the objective characteristics of human cognition itself, which could be unpacked by way of a dialectical process. Put simply, if Kant's project was to outline the subjective conditions of objective knowledge, then Hegel's project was to excavate the objective conditions of subjective knowledge.

Transferred to the matter of technology, Kantian technologies model aspects of a world beyond the inherent limits of the human senses, or a world inadequately available to the senses. These are technical devices that overcome the limits and failures of the senses—adjustments to telescope design that extend the failing eye gazing into chasmic space; dials in an airplane cockpit that guide the disoriented body in the sky; signal processing techniques that facilitate clear speech in contexts of whole-body vibration, such as spaceship launches; machine listening via acoustic sensors that track the sound of wind on Mars below the threshold of human audibility, and so on. The second set of technologies, construed here as Hegelian, model the inadequate senses *in themselves*. In this arena, the technical question is how to map the contingencies of the human sensorium in terms of a bitwise representation. How does a sound file, for example, preserve the fundamental acoustic fingerprints audible to the human ear of a sound in masked technical format? Is lossy compression, which irreversibly discards acoustic data to reduce file size, adequate to the purpose? The kinds of engineering that raise questions such as these work not toward attuning the human to the precision of a machine but toward constructing a machine that can ventriloquize the imprecision of the human sensorimotor apparatus itself. In other words, in today's computer–human interfaces, the human is not becoming virtualized as much as the machine is becoming humanized, approximating and mimicking the limits of the human ear.

What is fascinating about the technological turn to modeling human perception is that it offers both an illustration and a paradoxical inversion of the

Hegelian dialectical process. Hegel's dialectical process begins with a set of obvious claims about sense perception—or *sense certainty*—to reveal an understanding of these claims through an unfolding progression of determinate negations.[16] By tracing the dialectical progression through a series of contradictions and their supersession, Hegel's project culminates eventually in a kind of world explanation grasped as an objective reality or *concrete universal.* In contrast, by modeling human sensory experiences, the statistical turn in computing begins with a kind of world explanation—billions of data points readied for algorithmic processing—that reveals an understanding of the organic limits of sense perception. Sorting through a series of statistical correlations, AI culminates in a kind of human sensory certainty. While Hegelian in orientation, the machinic techniques AI deploys are a kind of dialectics in reverse—certainty without mediation, duplication without comprehension. To put this differently, epistemological inversion is what underwrites a technological shift away from perceptual augmentation toward perceptual modeling.

FINANCIALIZED PERCEPTION

The statistical imperative toward modeling human perception produces ontological commitments about both the nature of the human body in its sensorimotor workings and the phenomena with which the senses are engaging. By tracking human experience as material for datafication—or "rendering" experience as data—machine listening for music at once probes and encircles the limits of the human listening ear as well as the essential grounds of what constitutes music.[17] In other words, when a machine listens by way of statistically apprised predictive processing, it executes perceptual "absolutes"—or the absolute realization of perception in code.[18] In the next section, I address the potential limitations of these machine-inflected absolutes. Here I want to draw attention to the socioeconomic dimension of music informatics as a research tradition, taking note, in particular, of its imbrication in global capital markets. While research grants and official statements in the arena of MIR are often publicly grounded in inherent social values, such as alleviating disease and disability, their research outcomes are simultaneously attuned to market values and are frequently underwritten by commercialized digital networks. Communities of engineers, often assigned names such as the New Interfaces for Musical Expression community, collaboratively undertake technical research projects in academic contexts typically backed by public and private funding, effectively linking diverse stakeholders to research outcomes.

The Bregman Media Laboratories at Dartmouth College, for example, are sponsored by both corporate and government sectors. Their official logo, "Music, Mind, and Health," underscores public health as central to their mission. The computer scientists describe their project thus: "We are researching the mental representations underlying music perception and cognition as well as the mechanisms underlying musical affect. By developing new methods in computation and interpretation of neuro-imaging, this basic research lays the foundations for new audio-visual experiences as well as new applications to music therapies for health and healing."[19] The funding for these kinds of "making more health" projects comes from public institutions like the National Science Foundation and private ones like Google (now Alphabet). The director of Dartmouth's Bregman Media Labs, Michael Casey, received the *Google Faculty Research Award* in 2011 to support research on "Search by Groove," which I discuss in more detail later. The International Society of Music Information Retrieval is funded not only by educational institutions and members but by a host of corporate sponsors: Amazon Music, Bose, Gracenote, Pandora, Smule, Steinberg, Spotify, and Google. There are various advantages of the public–private funding model for MIR research. For example, some data sets are copyright protected, with fair use exemptions granted for educational and research purposes. Not-for-profit public institutions can thereby effectively clear the way for legal access to caches of data. They also contribute inexpensive labor to specific research projects and, as suggested earlier, function as an alibi for the social values served by the funding. However, as I show later, the economic value of this research—its commercial profitability—is deeply embedded in research outcomes. It would not be an exaggeration to speak of the research findings of machine listening as *sponsored* conceptual ontologies for perceptual technics, or more simply as *financialized perceptions.* What happens to perceptions that either fall outside the purview of the database or are somehow unusable, untagged, or uncommodifiable?

Recently, developers have been building algorithms for identifying the *groove* in music, software for instrument simulation or for style emulation, automatic *tempo* detection systems, algorithms for generalized *genre perception* or how humans *hear and imagine scales,* and automatic *tonal induction* and *beat tracking* software, among much else. In the case of style emulation, large collections of annotated musical data are used to train various deep neural network architectures with the aim of producing note sequences that both remain consistent to the musical style ascribed to that collection *and* present some form of novelty (known as the "creativity" of the built and trained network). Automatic compositional tools such as this are sometimes referred to as "creative-AI" and form a constitutive component of contemporary songwriting studios. Another example is digital groove tracking, an application that attempts to locate the so-called groove of music (its signature "feel" or sonic "vibe") in a machine-readable way. Michael Casey describes the process thus: "What if we could search large collections of music by groove? We want to understand the essence of what makes songs have a similar rhythmic feel. We will extract that essence using new machine hearing algorithms that we are developing for the project."[20] Historically speaking, the notion of a groove is not a traditional Western musical parameter like rhythm, meter, pitch, harmony, or even timbre; it is generally considered an elusive subjective affect, culturally localized and contested. The term itself—which can denote as little as a kind of feel-good phenomenon, such as "groovy" or "cool"—emerges from a particular technological moment in the history of music: the groove in a vinyl record. In other words, the groove in music is a modern idea, less quantifiable than a parameter like pitch or rhythm and frequently indexing phenomena outside of these traditional templates—the groove can be considered as the stretching or bending of metronomic time, for example, or certain characteristic modalities of repetition, or rhythmic groupings that cut against the grain of the music's metric beats. The software, however, deploys a digital detection scheme that statistically retrieves the groove in a user-friendly, human-centered environment. The machine thereby performs a kind of intelligent listening by processing voluminous data to realize typical musical capacities. To test results, researchers use functional magnetic resonance imaging (fMRI) to measure brain activity by detecting changes in blood flow of listening test subjects.

Apart from the obvious allegiance to a brand of statistically produced cognitivism, the very idea of a *groove* in music is modeled primarily on the recording practices and sonic stylistics of commercial popular music from the last fifty years. At the same time, with groove tracking software, music can be identified in ways that exceed both the limits of word-based queries (as found in Google-like search engines) and currently prevalent pattern-matching software, known as "audio fingerprinting" (as used in music search applications like Shazam). The software thereby contributes to new efficiencies in search functionality and algorithmic retrieval. For example, users can create queries by specifying a harmonic, melodic, or rhythmic constellation of notes by tapping a rhythm, humming a tune, singing, or whistling a melody.[21] The perceptually attuned machine listening system then retrieves a ranked list of possible candidates whose groove is musically related to the query. One might say that groove tracking does for the listening ear what the oracles of predictive text do for the typing hand; the software functions as an additional, artificial sense that sends envoys into the likely future by seeking out plausible forms—mined from statistical averages—from the recent past.

Again, decisions about the implementation and design of these technologies are largely made on the basis of efficiency and profitability. The commercial imperatives underwriting applications deploying groove recognition software include the identification of music for increased sales—as well as the identification of plagiarism of riffs, breaks, hooks, and feels—in ways more plastic than those permitted by the identical arrays of data, or pattern matching, in current recognition software. In this way, the software can increase profit through sales, leasing, and copyright enforcement. Recent litigation attests to the high commercial stakes for automatic groove tracking—a software that is capable of detecting a song's hook, feel, vibe, or genre. In 2015, for example, a court ruled that Robin Thicke and Pharrell Williams had infringed Marvin Gaye's copyright in their song "Blurred Lines" (2015), on grounds that it channeled the "feel" of Gaye's "Got to Give It Up." The testimony included a previous admission by Thicke in a *GQ* interview that he owed an allegiance to the "groove" of Gaye's music: "I was like, 'Damn, we should make something like that, something with that groove.'"[22] Although the ostensible composer of this groove operated in a much larger cultural milieu than the court admitted, Williams and Thicke were ordered to pay $7.4 million (later reduced to $5.3 million) to Gaye's surviving family. Likewise, in 2016, a women's rap group the Sequence sued Bruno Mars for lifting the *feel* of their song "Funk You Up" for the song "Uptown Funk." Legal claims of musical plagiarism on account of perceptual likeness are legion, but historically, they are rarely litigated on the terrain of *scènes à faire*. These are the generic stylistics—or genre-specific elements—of a song that have been legally construed as unprotectable ideas, rather than protectable musical expressions. Vesting ephemera like hooks, grooves, and feels with a kind of identifiable digital *materiality* may considerably expand what constitutes a tangible component of protectable musical expression.[23] In 2015 and 2016, however, cases like these were still litigated with considerable input from musical experts (at high expense), but they are coterminous with the development of the very idea of definitively locating and tracking a groove. Today, systems for automated copyright control, digital rights management, and other audio content detection systems are enhanced by groove trackers and are likely to play an increasing role in litigation as well.

On one hand, then, technologies like groove tracking promise to uncover plagiarism by automating an elusive similarity relation in music at much-reduced cost, bestowing upon a judgment of similarity an element of technical impartiality. On the other hand, groove tracking software can identify similarities that elude groove identity and thereby provide templates for creative near-imitations of rhythmic effects that would avoid the perils of copyright claims in a court of law. To this end, the software is a useful compositional tool for creative-AI. Furthermore, automatic groove tracking contributes to the formulation of enhanced, human-centered recommendation systems, operating on the basis of mobile autotranscription of gesture, voice, and so on. For example, applications that recognize songs on the basis of singing, whistling, or humming their main melody can be expanded to detect songs by way of a heard or felt groove. Song identification of this sort is frequently tethered to recommendations based on perceived similarity, which largely propels the algorithmically driven conveyer belt delivery systems for music today. A simple gesture or vocalization of a groove is thereby transformed into a list of songs on autoplay.

It is clear from this short set of applications that groove tracking technology is entangled with various practices—intersecting networks of technological, aesthetic, scientific, medical, economic, legal, and industrial agents. Stakeholders aside, it should also be added, this kind of software affords a genuine technosensory enrichment of sonic experience. Human interactions within the computer-mediated framework are experienced as more relevant, fluid, intense, flexible, and immediate than we find in a previous generation of applications. One may speak here of the "humanization" of the computer's temporal inputs to model biological perceptual faculties (thereby accommodating affective phenomena; nonobvious rhythmic periodicities; and, generally, more interactive musical encounters with a machine). Yet, the *modeling* of hearing is also a mode of *constructing* hearing; the felt interactivity and enhanced immediacy are underwritten by data processing methods for music that model and manipulate data according to a logic of large-scale statistical correlation as it intersects with apparently nonconscious biological intrusions of the body—changes in perspiration, breathing rates, blood flow, neural circuits, galvanic skin responses, and so on. While grounded in fixed technical standards and selective data averages, the software applications are paradoxically experienced as flexible and immediate, marking a broader shift today from what was once called psychopower to what is now called neuropower—a shift in this context from autonomous listening to the automation of listening.[24] For all their appearance as merely technical applications, lines of code are grounded in specific cultural interpretations. These predefined logics are not seen by users, and (while they exist in a parallel layer) appear decoupled from the resulting sounding forms. This software is the technical transcoding of listening according to rules and given data sets within a given style of audio projection. Software, one might say, operates as a kind of musical command system.

TECHNOPERCEPTUAL ENTRAINMENT

I will turn now to the second example of automatic music processing software, this time grounded in cognitive entrainment research. Known as beat tracking, this software detects temporal and structural regularities in musical inputs that coincide with the kinesthetically felt beats in musical listening. The capacity to track beats is a subset of a host of different types of biological entrainment, whereby the human body adjusts to various external cycles—diurnal patterns, tidal shifts, seasonal changes, hunger patterns, menstrual cycles, walking, running, and the like.

Where groove tracking is a modern idea, attendant to a narrow history of popular music (with its own history of engineering for profitable attention), beat tracking is understood to capture an archaic universal human trait known as beat induction. Beat induction is the cognitive skill that enjoins the human ear to perceive a regular pulse, which is then felt to persist within music.[25] In other words, the listening subject synchronizes to equidistant pulses, or projected pulses, extracted from acoustic inputs. Beat induction is considered by a cohort of cognitive psychologists and neuroscientists to be a fundamental musical trait—a domain- and species-specific skill that develops autogeneratively. Many commentators argue that this basic capacity played a decisive role in the origins of music, language, and so on.[26] Beat tracking software attempts to *model* the cognitive archetypes that underlie the human detection of periodicities in music. Its novelty rests in its ability to overcome the limits of quantization technologies like MIDI, whose discrete triggering signal representation schemes (mentioned earlier) inadequately map the more continuous perceptions of lived cycles of experience.

The first central challenge for beat tracking algorithms is that *felt* periodicities are not always reinforced by the sounding rhythmic activity of the actual music. In other words, beats in music can be felt without a corresponding sound to articulate them. As a result, listeners not only respond to salient time points in the musical flow but also anticipate them.[27] Such anticipatory attending to sonic phenomena cannot be correlated in a simple way to the peaks on a frequency curve of a spectrogram. Felt beats without actual rhythmic signals reinforcing them are called "metric accents" in music theory and sometimes dubbed "loud rests" by software engineers.[28] It is the task of the algorithmic beat tracker to hear the silent beat as metrically salient. Second, unlike the equidistant measures of clock or metronomic time, beat induction must consider the plasticity of the beat and its entrainment associated with musical time. In other words, human time tapping in music entails a degree of variability that is not perceptually registered as a variation of the time spans between beats. As it is with human beat induction, the listening computer should therefore be able to attribute equidistance to nonequidistant phenomena. Not surprisingly, music with sudden tempo changes additionally presents major obstacles for automatic beat trackers. Finally, as with most types of MIR, beat trackers need to overcome the challenges presented by vague (or noisy) sound signals.

Beat tracking software uses algorithms to extract psychologically perceived aspects of tempo, grouping, and meter from musical recordings or performances. As a first step, the tracking systems automatically extract salient events from the audio and convert the signal into feature representations. These features are extracted either from the signal's energy envelope (associated with perceptions of amplitude) or using a spectral-based approach (associated with perceptions of timbre or harmony). A derivative operator is then applied to the feature representation sequence to detect rhythmic onsets. The system then performs a periodicity analysis—a time–tempo representation based on recurring patterns—using Fourier analysis and autocorrelation techniques. This phase-based analysis thereby further differentiates the derived predominant local pulses into stronger and weaker beats. The earliest processing systems were grounded in rule-based ideas about meter formation, often inspired by linguistic grammars elaborated by Noam Chomsky and others.[29] In this literature, the perception of meter is said to be formed automatically according to certain cognitively hardwired metric preference rules (about which more later). The more recent turn to dynamic programming and recurrent neural networks—which track sequential features in a large data set to predict the probable next events in music—is fundamentally consistent with extant theories and studies that model hierarchical representations for rhythms.[30] Alongside the psychological modeling schema, beat tracking software today deploys the statistical analytics of big data sets and combines them with contemporary theories of brain functions.[31]

As with groove tracking, beat tracking software has a variety of commercial and noncommercial uses. Again, health, academic, and artistic criteria underwrite much of the research and its media representation. The work of Robert Bressir, for example, offers recuperative therapeutic strategies for neurogenetic diseases with the assistance of digital beat tracking software, while the research of Gerhard Widmer, Florian Krebs, and Sebastian Böck detects degrees of nonperiodic "swing" in classical music.[32] Most uses, however, have commercial value, such as music selection based on embodied beats (matching inhalation cycles, runners' steps, heart rates, and so on). It is noteworthy that the low-stakes disciplines of self-improvement, fitness, body modification, and health are a fertile ground for the instrumentalization of these noninstrumental sensory capacities. To take an elementary early example, in 2015, Spotify launched its Spotify Running product. In one of the first advertisements featuring a runner whose gait is matched by the sounding beat periodicities of electronic dance music (EDM), we find the contradictory claims—that this is "music that reacts to you," on one hand, *and* music that "let[s] the beat drive you," on the other—tersely intersecting an appeal to interaction and agency, on one hand, with interpellation and adaptation, on the other.[33] Spotify's step tracking analyzes the raw signal of the running body, filters the signal to extract the onset points and average tempo (or beats per minute) of the runner's steps, calculates a tempo estimation (somewhere between 140 and 200 bpm), and then launches a playlist or sound track that matches the tempo. The application provides nonstop beat matching between songs with DJ-quality transitions. The algorithm may also factor metrics such as the user's taste profile, default cultural community, time of day, weather, and geolocation. Spotify's elementary running app indicates an early example of automatic entrainment technology, but it does not yet activate the full functionality of recent beat trackers. These trackers are responsive to *changes* in tempo and time-stretching. In other words, the detected beats need not lock into a relatively stable average tempo, which is then matched to that of a given set of songs. Instead, in this new scenario, when the listening runner slows down or stops, the beat tracker could guide a transition toward a slower tempo or modulate toward ambient computer-generated music, effectively hitching the body to creative-AI.[34] In this way, the expanded functionality for beat tracking can be integrated into quotidian listener- and creator-oriented applications for sound tracks and playlists automated by behavioral informatics—tracking mood, time, movement, weather, activity, and location.[35]

That being said, most of the uses for beat trackers extend beyond everyday activity-oriented applications to a host of specialized uses. These include interactive technologies for real-time composition and performance, mixing and synchronization in cinema editing and postproduction, cover song detection in differently entrained biological temporalities, audio editing for digital DJs, and new music classification/browsing tools based on metric tempi.[36] The integration of beat trackers and accelerometers into full-body capture technologies—sometimes called interactive "wearables"—can generate sound tracks, compositions, and playlists that react in real time to networked subjects. Relevant streaming sensors monitor not only gestures and body motion but also social proclivities, environmental contexts, emotional states, ambient smells, or local weather conditions. Questions of pervasive surveillance aside, the point about this encroaching arena of biometric monitoring is that it is modeled on various specific theories of human perception that are easily and unobtrusively integrated into the low-stakes, opt-in environment of musical listening, creation, and interaction. Music informatics thereby shapes human sensing in a data-driven manner.

Cognitive theories regard the human sensory capacities as relatively invariant—instinctual, nonconscious, autogenerative, and so on—but the senses are also bound up in social forms and relations that cannot be simply subsumed into archetypal forms.[37] If animal attunement varies according to environmental coordinates, then human perception is bound up in, and metamorphosed by, social practices and technical interfaces. It is highly probable that the plasticity of perception involves complex noetic processes that cultivate technoperceptual life, if not the neurocortical organization of the brain itself. Algorithmic perception, then, is recoded by way of its technical, industrial, legal, and material mediation. Simply put, perceptual responses and affective behaviors to machinic inputs and outputs are less intentional or instinctual than they are a metastasization of a technical social process.

DEFAULT OCCIDENTALISM

It is beyond the scope of this essay to map all the varied input–output mechanisms in the context of these new interactive musical technologies. Yet it is instructive to take one of these mechanisms and examine the feedback loops its behavioral data encloses. To wit, what kind of technical social process is proffered by beat tracking? The best self-reported results for trackers emerge when they are set to work on popular music, notably rock, pop, and electronic dance music.[38] Furthermore, the tracker functions better in 4/4 and 2/4 time than it does in 3/4 time. It is also more accurately predictive for music in the region of 110 bpm and less so for music that is slow or tinged with expressive timing. In addition to industrial popular music, however, trackers produce quite successful results in the context of world music with a "high rhythmic character" or "clear percussive content."[39] When beat trackers are compared to control groups that tap out the beat to samples of African music, for example, we find a considerable overlap between human listening and machine listening.[40] This is a curious result given that data sets for commercial music applications tend to exclude African music. For example, the music-identification application Shazam includes only a tiny percentage of African music in its retrievable collection.[41] This is because Shazam's business model relies on licensed music, for which it is remunerated as it delivers data on searches to streaming services. In the words of Elena Razlagova, "Shazam shuns uncommodifiable scenes, because it has become a key player in the corporate music sphere. It has partnered with major online stores and streaming services—iTunes, Google Play, Spotify, Deezer, and others—that operate according to strict licensing restrictions."[42] As André Holzapfel and his collaborators attest, beat tracking technology, too, is most responsive to licensed industrial music.[43] It is thus surprising that control groups in laboratories tapping to African music tend to recapitulate machine listening.[44] The problem is that these control groups are infrequently located in the regions whose music is being tested. In the experiments conducted by Olmo Cornelis's team, the researchers were testing European subjects' perceptions of African music, which can render results in the image of Western popular music. For example, unaccompanied *mbira dza vadzimu* and *matepe* tunes of Zimbabwe are often tapped in duple meter by non-Shona listeners. Take a song like "Nhemamusasa," which, by contemporary Western theories of meter, is grouped in a duple 3/4 time.[45] This is presumably because listeners versed in Western classical and popular music follow either the harmonic changes or the bass movements of the interlocking texture—or what Lerdahl and Jackendoff would call the "metric preference rules" concerning harmonic rhythm and bass patterns, known as MPR 5.f and MPR 6, respectively.[46] It is not surprising that the current generation of beat trackers mimic rather precisely these apparently cognitively hardwired preference rules for modeling human perception.

And yet, the material interface of the *matepe* and *mbira* instruments proffers a set of efficiencies, functionalities, and limitations that program temporality and comport the body very differently than what is implied by either the automatic beat tracker or the control group in the Belgian laboratory. No less triggering-type musical *digitoria,* these African instruments are played in pairs, where one performer interlocks with—or plays within the spaces left by—the other. The woven arrangement produces a particular kind of ratchet-wheel aleatorics, which issues figures of asynchronous sound. Not only is the motor fact of the striking fingers delinked from the audible sound that comes to the ear but musical lines issue forth a kind of parallel polyphony of seemingly unplayed material—musical lines that escape the direct supervision of its producers. On the matter of locating the beat in this musical scenario, one can note several challenges. First, the interlocking players produce inherent patterns that transform with the smallest shift in motor pattern, relocating any audibly discernible beat thereby. Second, the inherent patterns resulting from the interlocking of the two parts are rhythmically irregular despite being produced by regular motor patterns that most obviously constitute beat formation. Third, metric entrainments produced by the inherent patterns are not aligned with the meter implied by the individual parts. Fourth, the variation techniques generally produce rhythmic motifs that are set adrift of the metric coordinates grounding them within any given time span.[47] In other words, the very conceptualization of rhythm and meter relations in African music could be revised in the context of the *mbira* or *matepe.* These are not isolated cases of musical types that resist the assumptions set forth by beat tracking. Similar arguments could be mounted across the terrain of numerous sub-Saharan African musics—the *amadinda* and *akadinda* music of Uganda, for example, or the *timbila* music of Mozambique, which offer equally striking Western meter–defying musical scenarios as the *mbira* or *matepe.*[48] Also, these are not simply cases of alternative modes of beat tracking but examples of musical scenarios that put the very idea of tracking a fixed beat into question. These precolonial African music temporalities set forth not mobile rhythmic figures on a fixed metric ground but fixed figures on a mobile ground. As a metric scenario, this kind of time shifting ensures that the music is maximally ambiguous. If metric entrainment is to be a central referent for musical listening at all, the African examples become a music to rotate the beat by—in short, an overwhelming challenge to dominant entrainment theories of metric perception.[49]

Of course, the digital beat tracker is hitched to a much simpler model of musical entrainment, grounded in popular music of the industrial Global North. Despite the largely successful overlap between human beat annotations and the machine listening of African music, these controlled tests cannot register the complexity of this African music. The software comports its listening schema away from the diversity of sensory behavior considered in a truly global frame and toward a generalized Eurogenetic industrial counterpart. The evolution of music software thereby contributes to a lengthy history of colonial imposition of cultural values in Africa and elsewhere. In his essay "Tonality as a Colonizing Force in Africa," for example, the Ghanaian musicologist Kofi Agawu details how—under the broad guise of a progressivist narrative—tonal musical thinking was exported into vast areas of Africa in the period of European colonialism. Often overlooked on account of its routine scope and authority in *post*-colonial Africa, the tonal system of pitch organization—"a *tertium quid* that integrates melody, harmony and metre into a single nexus"—became the dominant system for organizing musical life in African colonies, from Nigeria and Ghana in West Africa to Zimbabwe, Zambia, and South Africa in the south, by way of Kenya, Tanzania, and Uganda in the east.[50] In other words, tonal and metric schemes in African colonies were assimilated into European musical terms in the heyday of colonialism, bearing witness to the "gradual loss of a heritage language."[51] Likewise, in his *La Binarizacion de los Ritmos Ternarios Africanos en America Latina,* the Cuban musicologist Rolando Antonio Pérez Fernandez demonstrates how ternary time systems in African music gradually metamorphosed into binary ones in the context of the slave trade across the Atlantic ocean in the nineteenth century.[52] In the process of this transformation, many of the polyvalent metric properties of African music were neutralized and simplified. The contemporary automation of these dominant metric structures in code can be understood as a computational contribution to this legacy—a technical executor of trends that have the industrial backing to wash over musical life everywhere. The silent global dissemination of musical technologies such as this is likely to make its assumptions ubiquitous.

Yet, while relevant to a contemporary model of sense formation, these limiting algorithmic protocols are black-boxed for technical, legal, and financial reasons. First, lines of code exist in a kind of occluded dual setting, apart from the content they make visible and audible to the adopter; second, software is imbricated in networks of privacy agreements, licensing, patenting, and so on that generally keep it walled off from open adaptability; and third, given the orientation of its financial investments, the software is not engineered for a comprehensive accounting of the world's music, let alone accurately to represent uncommodified musical practices. My argument is that this software, by automating a particular practice of cultural mimesis, covertly remodels the perceptual life of individuals, leaving a kind of biological imprint on the body. Arguably, it is in this sense that software is actually *harder* than hardware, for it is less immaterial or malleable than it is materializing a perceptual absolute.[53] Through a series of meticulous symbolic declarations and descriptions, it *exteriorizes* what was once an inner automatism in a financially viable sense.[54] When software models human perception, it executes ontologies (or absolutes) more than it builds concepts (or interpretations). The software for music is, in this fundamental sense, the opposite of the hermeneutics of music.

By layering statistically derived axioms of the listening ear into these lines of code, musical interaction enters into a feedback loop that is equally a site of interpellation. Once the software is widely adopted in practice, content provision of streaming media gradually modifies the basic nature of musical listening. The subjective experience of interactive applications that model human perception in a sensuous, immediate, and intuitive way silently forecloses what goes as human time tapping, the *feel* of sounds, musical entrainment, and perhaps even time itself. The technical curation of musical taste, one might say, becomes a curation of musical perception itself. Enmeshed as they are in the commercial networks that underwrite their development, beat trackers—no less than groove detection technologies, models for instrument simulation, automatic style emulation, and so on—quietly circumscribe human perception within a localized, arguably contingent framework. Has music informatics become an instrument for a programmed cognitivism profitably instructed by specific financial imperatives? Would it be an exaggeration to speak of a default industrial Occidentalism, a perceptual entrainment algorithmically tethered to the spirit of capital?

NOTES

I would like to thank Sarah Wasserman and the editorial team for their attentive engagement with this chapter, no less than for their encouragement. I am additionally grateful to Kofi Agawu, Andre Holzapfel, Tristan Jehan, and Viktoria Tkaczyk for their astute interlocution over the years. This chapter would not have been possible without their input.

1 On music's ineffability, see Carolyn Abbate, "Music: Drastic or Gnostic?," *Critical Inquiry* 30 (2004): 505–36, and Michael Gallope, *Deep Refrains: Music, Philosophy, and the Ineffable* (Chicago: University of Chicago Press, 2017).

2 For an introduction to these basic concepts, see Tristan Jehan's "Creating Music by Listening," PhD diss., MIT, 2005.

3 Tristan Jehan, Paul Lamere, and Brian Whitman, "Music Retrieval from Everything," paper presented at the 11th ACM SIGMM International Conference on Multimedia Information Retrieval, 2015.

4 For a recent example of the way musical assumptions were freighted into the African colonies, see Ronald Michael Radano and Tejumola Olaniyan, eds., *Audible Empire* (Durham, N.C.: Duke University Press, 2016).

5 It is beyond the scope of the essay to historicize the various technological, aesthetic, social, scientific, political, and industrial contingencies underlying the construction of even the most friction-free (or natural-seeming) applications for automated machine listening. Historical analyses are themselves the result of a vexing network of competing views and interests. In other words, a historical engagement with software applications for machine listening in music entails, at a minimum, a history of the software industry itself and, in particular, a history of various relevant algorithmic systems and designs derived from mathematics—from the eighteenth-century Markov model (itself a subset of probability statistics) to the nineteenth-century Fourier transform (itself a subset of frequency signal analysis), and so on. Simultaneously, historical engagement with these applications entails, at a minimum, a history of music cognition and music theory, competing theories pertaining to meter and rhythm, and even divergent conceptualizations of musical temporality itself. None of these concepts and theories are stable over time, and yet the fundamental operation of machine listening—paradoxically tethered to practically unlimited data sets—coheres around relatively stable or typical statistical entities. Further historical inquiries into the question of machine listening include a history of technical standards (from standardized pitch tuning to metric schemes, now commonly adopted as points of reference for musicians in the industrialized world); a history of musical automata (mechanisms, from ancient music boxes to modern piano rolls, designed automatically to carry out predetermined operational sequences or instructions); and a history of time

warping itself (from ancient attempts to bring astronomical periodicities into conformity with variously projected calendric temporalities to the widespread modern philosophical inflection of being with time, construed as somehow outside of measured time). The various avenues for historical inquiry could be multiplied, intersected, and diversified, each avenue subtly inflecting the historical picture.

6 Fred Lerdahl and Ray Jackendoff, *A Generative Theory of Tonal Music* (Cambridge, Mass.: MIT Press, 1983).

7 John Haugeland, *Artificial Intelligence: The Very Idea* (Cambridge, Mass.: MIT Press, 1985).

8 For a detailed account of how data-driven machine learning superseded traditional linguistic modeling in the context of predictive text and speech recognition systems, see Xiaochang Li, "Divination Engines: A Media History of Text Prediction," PhD diss., New York University, 2018. For a brief introduction to the distinction between human intuition and machinic pattern recognition, see Clive Thompson, "The Miseducation of Artificial Intelligence," *Wired,* December 2018, 74–81.

9 Frederick Jelinek, quoted in Peter Hillyer, "Talking to Terminals," *THINK,* 1987, IBM Corporate Archives.

10 For a useful short history of the MIDI protocol, see Ryan Alexander Diduck's "The 30th Anniversary of MIDI: A Protocol Three Decades On," *Quietus,* January 22, 2013, https://thequietus.com/articles/11189-midi-30th-anniversary.

11 On the event-based mosaic templates of MIDI, see Nicola Bernardini and Gareth Loy's "Whither MIDI?," *Computer Music Journal* 11, no. 1 (1987): 9–12. On the problematics of unpredictable microtiming in MIDI-enabled performances, see Richard F. Moore's "The Dysfunctions of MIDI," *Computer Music Journal* 12, no. 1 (1988): 19–28. On the limits of unidirectional communication, see Gareth Loy's "Musicians Make a Standard: The MIDI Phenomenon," *Computer Music Journal* 9, no. 4 (1985): 8–26. These concerns constitute a small subset of characteristic discontents voiced about MIDI in the 1980s. More recent critical views include Jon Drummond's "Understanding Interactive Systems," *Organized Sound* 14, no. 2 (2009): 124–33, and Jaron Lanier's *You Are Not a Gadget* (New York: Vintage, 2010).

12 For an eloquent account of this interest in non-MIDI-based music making with machines, see David Wessel's "Designing Musical Instruments That Privilege Improvisation," Center for Interdisciplinary Research in Music Media and Technology (CIRMMT), Distinguished Lecture in the Science and Technology of Music, November 25, 2010.

13 On the Million Song Dataset, see Thierry Bertin-Mahieuz, Daniel P. W. Ellis, Brian Whitman, and Paul Lamere's "The Million Song Dataset," paper presented at the 12th International Society for Music Information Retrieval Conference,

ISMIR, 2011, http://ismir2011.ismir.net/papers/OS6-1.pdf. On the Music Genome Project, see the short historical statement by Tim Westergren, "The Music Genome Project," Pandora website, http://www.pandora.com/about/mgp.

14 Joseph C. R. Licklider, "Man–Computer Symbiosis," *Transactions on Human Factors in Electronics* HFE-1 (March 1960): 4–11.

15 Joseph Grudin, "AI and HCI: Two Fields Divided by a Common Focus," *AI Magazine,* 2009, 48–57. Likewise, technology and science reporter for the *New York Times,* John Markoff, makes a distinction between intelligence augmentation (IA) and artificial intelligence (AI): one research tradition aims to "extend human capacities," whereas the other attempts to "replace human beings with intelligent machines." Markoff, *Machines of Loving Grace: The Quest for Common Ground between Humans and Robots* (New York: Ecco Books, 2015), xii.

16 On sense certainty, see the opening arguments of Georg Wilhelm Friedrich Hegel, *Phenomenology of Spirit,* trans. A. V. Miller with a foreword by J. N. Findlay (Oxford: Clarendon Press, 1977).

17 For a comprehensive account of "body rendition" in the context of surveillance technologies, see Shoshana Zuboff, *The Age of Surveillance Capitalism: The Fight for a Human Future at the New Frontier of Power* (London: Profile Books, 2019), 242–54.

18 This position takes its cue from Alexander Galloway, who argues that the mediated simulation is a purely ideological form. Galloway writes, "The virtual can only exist within the absolute; the virtual needs the absolute. Yet conventional wisdom often suggests the reverse, that the virtual is the thing that stands 'above' or apart from the real, that all anxieties about the real ultimately find their escape in the virtual. But here the conventional wisdom is wrong, for the exact opposite is true. The virtual can only be possible, not in relation to the real, but in relation to the absolute." See Galloway, *The Interface Effect* (Cambridge: Polity Press, 2012), 124.

19 For the official mission statements of the Bregman Media Labs, see http://bregman.dartmouth.edu/.

20 Michael Casey, quoted in Kelly Sundberg Seaman's "Dartmouth Professor of Music Awarded from Google," June 9, 2011, https://news.dartmouth.edu/news/2011/06/dartmouth-professor-music-awarded-grant-google.

21 In late 2020, various applications were released that deployed this kind of intuitive feature for finding songs. For example, on Google's search widget, the "hum to search" feature allows users to hum, sing, or whistle a melody to locate songs. See Chaim Gartenberg's "Google's New 'Hum to Search' Feature Can Figure Out the Song That's Stuck in Your Head," *The Verge,* October 15, 2020, https://www.theverge.com/2020/10/15/21518182/google-new-hum-to-search-feature-identify-song-machine-learning.

22 Robin Thicke, quoted in Ben Beaumont-Thomas's "Robin Thicke and Pharrell

Williams to Pay $5m in Final Blurred Lines Verdict," *Guardian*, December 13, 2018, https://www.theguardian.com/music/2018/dec/13/robin-thicke-and-pharrell-williams-to-pay-5m-in-final-blurred-lines-verdict.

23 It is instructive to consider the effects of the digitization of culture less in terms of the dematerialization of tangible media than its dramatic rematerialization. After the landmark case *Bridgeport Music v. Dimension Films* in 2005, for example, a technically manipulated sample (bearing no *perceptible* likeness to the source) could nonetheless be construed as infringing a legal right on the basis of its "physical" contact with a "fixed medium." The case effectively created a precedent for the elimination of the de minimus doctrine for sampling recorded music. Given that the court's decision in *Williams v. Gaye* in 2017 grants the "feel" and "groove" of a song legal protection, groove trackers (as with audio samples in *Bridgeport*) can transform an undefinable sonic property into a tangible entity or sonic signature. The case creates a precedent for the elimination of heretofore unprotectible ideas, or *scènes à faire*.

24 On the concept of "neuropower," see Bernard Stiegler, "Social Networking and Neuropower," presentation at New York University, April 13, 2013.

25 For a useful introduction to the biomusicological concept of beat induction, see P. Desain and H. Honing's "Foot-Tapping: a Brief Introduction to Beat Induction," in *Proceedings of the 1994 International Computer Music Conference* (San Francisco: International Computer Music Association, 1994), 78–79.

26 While the literature on musical entrainment and the origins of music in entrainment is vast, some representative recent texts include Jenny Judge's "'Feeling the Beat': Multimodal Perception and the Experience of Musical Movement," Michael Spitzer's "Mozart's 'Dissonance' and the Dialectic of Language and Thought in Classical Theories of Rhythm," and "Metric Entrainment and the Problem(s) of Perception," all in Peter Cheyne, Andy Hamilton, and Max Paddison, eds., *The Philosophy of Rhythm: Aesthetics, Music, Poetics* (Oxford: Oxford University Press, 2019), 76–90, 125–40, and 171–82, resp.

27 On the nature of temporal attention to stimuli, see M. R. Jones, H. Moynihan, N. MacKenzie, and J. Puente, "Temporal Aspects of Stimulus-Driven Attending in Dynamic Arrays," *Psychological Science* 13 (2002): 313–19.

28 On the various kinds of accent in music, see Lerdahl and Jackendoff, *A Generative Theory of Tonal Music*, 17.

29 For a description of rule-governed expert systems for beat induction, see P. Desain and H. Honing, "Computational Models of Beat Induction: The Rule-Based Approach," *New Music Research* 28, no. 1 (1999): 29–42. For an example of a rule-governed expert system, see H. Christopher Longuet-Higgins and Christopher S. Lee, "The Perception of Musical Rhythms," *Perception* 11, no. 2 (1982): 115–28.

30 On beat tracking with recurrent neural networks, see Sebastian Böck, Florian Krebs, and Gerhard Widmer, "Joint Beat and Downbeat Tracking with Recurrent

Neural Networks," paper presented at the 17th International Society for Music Information Retrieval Conference (ISMIR), 2016.

31 See, e.g., E. Basar, "Toward a Physical Approach to Integrative Physiology: Brain Dynamics and Physical Causality," *American Journal of Physiology* 245, no. 4 (1983): 510–33, and P. Nunez, *Electric Fields of the Brain* (New York: Oxford University Press, 1981).

32 See, e.g., Sebastian Böck, Florian Krebs, and Gerhard Widmer, "A Multi-model Approach to Beat Tracking Considering Heterogeneous Musical Styles," paper presented at the 15th International Society for Music Information Retrieval Conference (ISMIR), 2014, and Böck, Krebs, and Widmer, "An Efficient State-Space Model for Joint Tempo and Beat Tracking," paper presented at the 16th International Society for Music Information Retrieval Conference (ISMIR), 2015.

33 See https://www.youtube.com/watch?v=4-7IZzcpz8o and https://www.youtube.com/results?search_query=spotify+running+ad+2015.

34 In late 2020, Weav Music, an "adaptive" music application founded by two computer scientists and software developers Lars Rasmussen and Elomida Visviki, integrated beat trackers into an application that could generate playlists that seamlessly altered the music's tempo, mood, and form according to the movements, activities, and geolocation of listeners. Rasmussen's background in mapping-related software at Google and "semantic search" algorithms at Facebook laid the groundwork for this flexible mode of automated playlisting. See https://www.weav.io/.

35 This new type of playlist needs to be contextualized in relation to financial imperatives. As music streaming companies, such as Spotify, begin to create, curate, and mount their own content, they can circumvent licensing and copyright costs of independent rights holders. This content is most readily integrated into mood- and activity-based playlists (rather than artist-based ones) and can be considerably enriched by the perceptually attuned technologies discussed earlier. See Tim Ingham, "Spotify Is Making Its Own Records . . . and Putting Them on Playlists," *Music Business World*, August 31, 2016, https://www.musicbusinessworldwide.com/spotify-is-creating-its-own-recordings-and-putting-them-on-playlists/.

36 Prominent figures in the field of beat tracking in music information retrieval include Andre Holzapfel, Fabien Guoyon, Annsi P. Klapuri, Matthew E. P. Davies, José R. Zapata, and João Lobato Oliveira, among many others. For an excellent overview of the technical dimensions of beat tracking, see Meinard Müller, *Fundamentals of Music Processing: Audio Analysis, Algorithms, Applications* (Cham, Switzerland: Springer, 2015), 303–46.

37 An exemplary case of music theory that construes meter formation in terms of invariant cognitive archetypes (known as metric preference rules) is Lerdahl and Jackendoff, *A Generative Theory of Tonal Music.* The book is frequently cited in the literature on beat tracking.

38 See André Holzapfel, Matthew E. P. Davies, José R. Zapata, João Lobato Oliveira, and Fabien Gouyon, "Selective Sampling for Beat Tracking Evaluation," *IEEE Transactions on Audio, Speech, and Language Processing* 20, no. 9 (2012): 2539–48. The paper demonstrates the musical properties "where future advances in beat tracking research would be most profitable and where beat tracking is too difficult to be attempted," noting that "rock, pop and electronic dance" genres are best suited to beat tracking algorithms (2539).

39 Fabien Gouyon, personal conversation, 2015, and Holzapfel et al., "Selective Sampling," 2539.

40 See, e.g., Olmo Cornelis, Joren Six, André Holzapfel, and Marc Leman, "Evaluation and Recommendation of Pulse and Tempo Annotation in Ethnic Music," *Journal of New Music Research* 42, no. 2 (2013): 131–49. By comparing automatic with human annotations, the paper analyzes automated tempo estimation (which signals beat formation) in the context of African music.

41 On the absence of African music in Shazam's database, see Elena Razlagova's "Shazam: The Blind Spots of Algorithmic Recognition and Recommendation," in *Appified: Culture in the Age of Apps*, 357–66 (Ann Arbor: University of Michigan Press, 2019).

42 Razlagova, 261–62.

43 Holzapfel et al., "Selective Sampling."

44 Cornelis et al., "Evaluation and Recommendation."

45 For a simple introduction to this complex discussion, it is instructive to compare Paul Berliner's iconic recording of "Nhemamusasa" ("Nhemamusasa: Complete Performance," YouTube video, 0:15, October 17, 2015, https://www.youtube.com/watch?v=ZBr5eohenIc) with the appropriation of "Nhemamusasa" by the Penguin Café Orchestra in their song "Cutting Branches for a Temporary Shelter" ("Cutting Branches for a Temporary Shelter—Penguin Cafe Orchestra," YouTube video, 3:10, January 4, 2012, https://www.youtube.com/watch?v=xYUAudoC-2A). Although the Penguin Café Orchestra's 1988 rendition is a note-for-note rendition (likely taken from Berliner's 1980 book *The Soul of Mbira*, given the resonance of the title with the Shona translation of the tune), the musicians interpret the music in a binary 3/4 instead of a ternary 12/8 time. "Cutting Branches" is rendered in a plain and folkish manner on an equally tempered pianica. Instead of interlocking with a second *kutsinhira* part, the Penguin Café Orchestra's version elaborates an unvarying *kushaura* part with some simple upper register variations. Without judging the (de)merits of the appropriated tune, the American musicians—deeply consistent with the metric entrainment schemes of Euro-industrial popular music—hear the Zimbabwean music in a sequence of beats organized according to a binary schema in triple time, instead of a ternary one in quadruple time. This same tune is featured in a Google Doodle in 2020 (ostensibly representing the "ancient instrument of the Shona people") and prominently exhibits a similar

perceptual ambiguity between two modes of metric entrainment, exemplified by the music heard before and after the entry of the *hosho* (rattle) part (https://www.google.com/logos/2020/mbira/r4/mbira20.html).

46 Lerdahl and Jackendoff, *A Generative Theory of Tonal Music,* 84, 88.

47 For a detailed account of the metric ambiguities of *mbira* and *matepe* music, see my "Temporal Geometries of an African Music," *Music Theory Online* 16, no. 4 (2010), https://www.mtosmt.org/issues/mto.10.16.4/mto.10.16.4.scherzinger.html.

48 On the formation of inherent patterns in Ugandan xylophone music, see Gerhard Kubik's *Theory of African Music,* vol. 1 (Chicago: University of Chicago Press, 2010); Kubik's "The Structure of Kiganda Xylophone Music," *African Music Society Journal* 2, no. 3 (1960): 6–30; and Kubik's "The Phenomenon of Inherent Rhythms in East and Central African Instrumental Music," *African Music* 3, no. 1 (1962): 33–42. On the rhythmic-metric complexities of the xylophone music of Mozambique, see Andrew Tracey's "Chopi Timbila Music," *African Music: Journal of the International Library of African Music* 9, no. 1 (2011): 7–32.

49 Some ethnomusicologists have attempted to describe the unique character of African musical temporalities in these terms. See John Miller Chernoff, *African Rhythm and African Sensibility: Aesthetics and Social Action in African Musical Idioms* (Chicago: University of Chicago Press, 1979). Given the sparse conditions of the archive, however, we find very few genuine music theories pertaining to precolonial African music. The writings of, among others, Gerhard Kubik, Akin Euba, Andrew Tracey, Kofi Agawu, Meki Nzewi, Kwabena Nketia, Sylvia Tamusuza, and David Locke could be considered a contribution to such a project.

50 Agawu, "Tonality as a Colonizing Force in Africa," in *Audible Empire,* ed. Ronald Michael Radano and Tejumola Olaniyan (Durham, N.C.: Duke University Press, 2016), 335.

51 Agawu, 351.

52 Rolando Antonio Pérez Fernandez, *La Binarizacion de los Ritmos Ternarios Africanos en America Latina* (Havana: Casa de las Américas, 1956).

53 On the emergence of the "absolute" in the context of mediated simulation, see Galloway, *Interface Effect.*

54 On the idea of an inner automatism, see Theodor W. Adorno, *Current of Music,* trans. Robert Hullot-Kentor (Cambridge: Polity, 2009), 109.

Part III

Making

6

The Useful Arts of Nineteenth-Century Patent Models

REED GOCHBERG

On December 28, 1840, the U.S. Patent Office issued Patent No. 1,915 to Rufus Porter for a fire alarm. Along with a written description and a drawing of the invention, Porter submitted a small model, consisting of a square wooden box with a bell and mechanism inside and an elaborate painting on its exterior. Complete with a simple gilt frame, the painting draws the eye to a dramatic scene of a small house on fire, with flames shooting out of the roof and windows and people rushing to and from the blaze. Beneath the scene, large lettering announces "The Sensitive Fire-Alarm. Invented by Rufus Porter, Billerica, Mass. 1840" (Figure 6.1).

Porter's patent application represented his invention in three ways: a written specification, a drawing, and this model, each of which was officially intended to showcase the novelty of his idea and demonstrate how it worked to an audience of patent examiners and specialists. The specification and drawing focus on the improvements he had made to the internal mechanism of his fire alarm, revealing how heat and smoke would trigger a bell inside the wooden box. The patent model, however, attracts the eye to the box's exterior, where the painted scene illustrates the potential stakes of Porter's invention. While this decorative facade served no functional purpose, it allowed Porter to make his idea more obvious to viewers who might otherwise confront a fairly mundane wooden box. The painted exterior immediately identifies this object as a fire alarm, adding visual appeal to an otherwise hidden mechanism.

Throughout the nineteenth century, thousands of patent models were submitted by inventors, from steam engines and industrial machinery to toys, measuring tools, and household implements. Depending on the invention, the materials used might vary, yet inventors frequently included some form of decorative flourish, even if it was merely to paint a wood or tin model to more closely resemble its imagined full-sized counterpart. By focusing in detail on a few individual examples of patent models, this essay aims to illuminate the strategies nineteenth-century inventors

FIGURE 6.1. Patent model, fire alarm. Rufus Porter, December 28, 1840. Hagley Museum and Library.

used to create patent models that might attract broad audiences. Through advertisements, bold ornaments, and juxtapositions of texture and materials, inventors sought to position their work to attract potential manufacturers and consumers. These seemingly irrelevant and even distracting features illuminate practices of making and modeling through which inventors combined specialized knowledge, familiar elements, and visual appeal to make their work accessible and engaging to the broadest possible audience.

These features, however, also occasionally disrupted the relationship between the model and the invention it promised to represent. Like the other components of a patent application, the model's purported audience was the examiners and clerks at the U.S. Patent Office, who could use three-dimensional models to move individual parts, test a hinge or bolt, or otherwise reveal a working mechanism at scale. Yet while the specification and drawing would use technical language to outline the mechanical details, inventors took much more creative liberties in crafting their models. A drawing might illustrate the single individual mechanism being patented, but the model frequently represented the entire machine, blurring the lines between part and whole and emphasizing the later product that might result from such an improvement. In a series of essays on a theory of signs, Charles Peirce defines icons as "likenesses," "which serve to convey ideas of the things they represent simply by imitating them."[1] By using patent models to provide "likenesses" of a later product, inventors not only unsettled the mimetic relationship between model and patent application but also disrupted the bureaucratic purpose of these objects and their broader goals for crafting them.

By examining Porter's fire alarm alongside other examples of nineteenth-century patent models, we can see how inventors played on the possibilities of representation that were afforded by three-dimensional modeling. Historians of science have examined the role of models in bridging the divide between specialists and broader audiences through pedagogy, demonstrations, and display, suggesting how models can reveal the broader social and cultural contexts that surround the production of knowledge in science and technology.[2] Such contexts are particularly visible in the decorative features attached to patent models, which highlight the links between the material culture of scientific practice, consumer culture, and the domestic sphere.[3] Although legal historians have examined patent models in relation to the history of intellectual property and the development of the patent system, these objects also open up possibilities for understanding the close proximity between material culture, decorative arts, and technological innovation.[4] Studies of digital media have emphasized the interfaces and other forms of mediation that structure engagement with new technology, and Johanna Drucker warns against "blindness to the rhetorical effects of design *as a form of mediation* (not of transmission or delivery)."[5] Making inventions more broadly legible involved such processes of mediation that restructured the kinds of information embedded in these objects. As the examples of a fire alarm, lightning rod, rocking chair, and parlor stove will demonstrate, patent models combined mechanical functionality with aesthetic appeal to situate new technology in the familiar context of the middle-class home.

The most visually appealing aspects of patent models—their bright ornaments, colorful advertisements, and miniature details—were also the elements that were least necessary to their stated purpose. When we examine these features more closely, however, we can see more clearly how inventors understood the importance of legibility and familiarity to conveying the potential value of their work. The strange and even surprising elements of patent models were not incidental to their mechanical functions; rather, they played an essential role in inviting viewers to imagine the aesthetic and affective potential of new technology.

REPRESENTING INVENTION

Only a brief period in the history of American innovation can be charted through patent models. While many inventors had voluntarily included models of their inventions since the establishment of the Patent Office in 1790, the Patent Act of 1836 formalized the model requirement, asking inventors to submit three forms of representation: a written specification, a drawing, and "a model of his invention, in all cases which admit of a representation by model, of a convenient size."[6] The Patent Act outlined the distinct role that each form of representation was intended to play. The specification would "enable any person skilled in the art or science to which it appertains, or with which it is most nearly connected, to make, construct, compound, and use the same" and "particularly specify and point out the part, improvement, or combination, which he claims as his own invention or discovery." Similarly, drawings and "specimens of ingredients" would guide patent examiners in "the purpose of experiment."[7] Finally, a model of the invention would "exhibit advantageously its several parts."[8] Through these three forms of representation—textual, visual, and material—an inventor could establish a claim to a new idea and acquire a limited-term monopoly on its applications in manufacture and production.

These new requirements accompanied the expansion of the bureaucracy that surrounded the patent system, including hiring more examiners and clerks, appointing a new commissioner of patents, and designing an imposing building that would be completed in 1840. Over the course of several decades, the halls of the Patent Office became increasingly crowded with models. The half-century period when models were a requirement was bookended by two fires, one in 1836, which destroyed some of the Patent Office's early collections, and one in 1873, in which seventy-five thousand out of an estimated two hundred thousand models were destroyed.[9] In 1880, the government abolished the requirement because of a lack of space, and in the early twentieth century, Smithsonian curators decided to keep only a selected number of patent models and discard the rest. Nonetheless, the thousands of models that survive at the Smithsonian, in the Hagley Museum and Library, and in private collections cover a variety of inventions, from industrial machinery, such as the power loom and steam engine, to household tools, such as washing machines, stoves, and furniture. The models could measure up to twelve cubic inches, but they frequently varied in size, scale, and materials. While many model machines were constructed of wood, steel, or brass to demonstrate their mechanisms, other patent models included tin, paint, and even fabric.[10] As a result, models ranged widely both in type of invention and form of composition.

Patent models were officially required to represent how individual inventions worked for patent examiners, clerks, and judges. The U.S. Constitution established patent law "to promote the progress of science and useful arts," and such "useful arts" could include ideas and processes as well as mechanical or technological innovations, as long as the invention was both novel and useful. As three-dimensional representations of intangible ideas, patent models were used in both the bureaucratic and legal spheres to adjudicate between competing claims to novelty and utility. Alain Pottage and Brad Sherman have suggested how patent models reinforced broader cultural associations between patents and machines. In a courtroom setting, they functioned as scale models, showing how a redesigned or improved feature operated. Describing the "figures of invention" created within the system of patent law, Pottage and Sherman have argued that these forms of representation helped to shape broader conceptions of invention as a material practice, as opposed to an intangible idea: "The intangible is only visible in the material shape, configuration, and operation of a material artefact or process where it reveals itself."[11] As a result, they argued, patent models could be interpreted as both idea and embodiment, designed to represent an intangible concept in material form. In this way, patent models expanded on the kinds of representation enabled by the specification and drawing, enabling inventors to demonstrate how an idea described on paper functioned in space.

While patent models were intended to show that an idea worked in practice, the written specification provided more explicit details of how an invention was novel, useful, and therefore patentable. In his 1849 treatise on patent law, George Ticknor Curtis emphasized the importance of a clear written description for articulating an inventor's contribution: "If the patentee has left it wholly ambiguous and uncertain, so loosely defined, and so inaccurately expressed, that the court cannot, upon fair interpretation of the words, and without resorting to mere vague conjecture of intention, gather what the invention is, then the patent is void for this defect."[12] While Curtis acknowledged the potential benefit in obscuring the exact means of replicating the invention, he argued for clear and descriptive prose to ensure the patent's approval.

These requirements demonstrate the various kinds of skills inventors had to possess to present their patent applications for approval by the federal government and to market them to broader audiences of manufacturers and consumers. During this period, the patent system gradually expanded to include a range of supporting enterprises, including a contingent of patent clerks and examiners who reviewed applications in Washington, D.C.; publications such as *Scientific American* and its affiliate patent agency Munn & Co., which assisted inventors with their applications; and a gallery to house and display models at the U.S. Patent Office.[13] The gallery attracted thousands of visitors and tourists every year, who frequently recorded their impressions of the many crowded cases of patent models in newspapers and travel narratives.[14] Inventors also often included their names and addresses on their patent models, perhaps as a practical strategy to ensure their models were correctly attributed, but these eye-catching advertisements also suggest their awareness of the possibilities for self-promotion and marketing. Scholars have emphasized the increasing professionalization of invention during the nineteenth century, indicated by such expanded federal bureaucracy and the marked increase in the number of patent agents and attorneys.[15] Patent specifications and drawings described inventions in technical language that was clearly aimed at an audience of specialists. Some models similarly emphasized only the technical features of an improvement, but many others suggest how inventors saw the advantages in promoting their work to numerous audiences at once.

Despite the protocols of the Patent Office, however, patent models revealed an ongoing playfulness when it came to three-dimensional representation. If documents such as the specification and drawing, as Curtis suggested, left no room for "ambiguous and uncertain" renderings of the invention, patent models enabled inventors to represent the whole, rather than an individual part, complementing these other forms of description even if not directly aligning with them. The presence of the specification and drawing also allowed inventors to take more creative liberties in experimenting with the form of the patent model. Even as these objects played a significant role in the legal and bureaucratic aspects of the patent process, they also opened up possibilities for representing mechanical improvements in more aesthetically appealing ways.

USEFUL ARTS

Rufus Porter's fire alarm reveals the challenge inventors faced in attempting to represent small or even mundane improvements to existing machines. Most patents were issued for relatively minor improvements: a modification to a bolt, screw, or hinge; a different mechanism to improve efficiency; or a new design for an existing machine. While Porter's model adheres to the official, professional standards for meeting a legal requirement, its most eye-catching feature—the painting on its exterior—creates distance between the model and its attached specification and drawing by concealing its patentable features. By including such extraneous elements, Porter was not only making his patent model more visually exciting; he was also distracting viewers from the most valuable information being represented.

The painting on the exterior of Porter's fire alarm resonates with his long-standing interest in promoting the spread of artistic and scientific knowledge. In his early life, Porter worked as an itinerant painter, traveling New England painting portraits and landscape murals on the walls of homes and inns. Like his fire alarm, these scenes frequently featured a multistory house surrounded by trees and a river in similarly flat perspective. In an instructional manual entitled *A Select Collection of Valuable and Curious Arts, and Interesting Experiments* (1825), Porter assembled a series of instructions for mixing paint colors, applying paint on numerous surfaces, and painting landscapes. In the "Advertisement" to *Curious Arts,* he describes painting through scientific method, promising "a more accurate and extensive knowledge of those arts" and "chemical experiments" that will "combine recreation with improvement in useful knowledge."[16] In his "recipes" for mixing paint colors, Porter emphasizes the chemical processes that produce the aesthetic effects of individual colors. Additionally, he includes specific suggestions as to color, proportion, and size, offering his readers a manual to reproduce his images and create their own. His early writings on landscape painting anticipate his later work as an inventor and editor, presenting ideas for "improvements" to the methodology of creating works of art.

While Porter's early writings offered methods in the decorative arts, his later career emphasized the popularization of scientific knowledge. After receiving the patent model for the fire alarm, Porter himself would continue to propose then-outlandish ideas for inventions, most famously a series of drawings and models of an "airship," a flying machine that he exhibited to fascinated audiences in New York and Washington, D.C.[17] Over the next few decades, he received several patents for a life preserver, a floating dock, and other inventions. Beginning in 1845, Porter established and edited *Scientific American,* a widely circulated journal that played a significant role in sharing ideas about new inventions and the patent system. In the magazine's first issue, published August 28, 1845, Porter announced its wide-ranging interests, including

> general notices of the progress of Mechanical and other *Scientific Improvements*; American and Foreign Improvements and Inventions; Catalogues of American Patents; Scientific Essays, illustrative of the principles of the sciences of Mechanics, Chemistry and Architecture; useful information and instruction in various Arts and Trades; Curious Philosophical Experiments; Miscellaneous Intelligence, Music and Poetry.[18]

By publishing accounts of new and improved inventions, advice about navigating the patent system, and other general interest articles on chemistry and mechanics, Porter implicitly promised to transform his readers into inventors. While the magazine announced a wide range of interests, including "Music and Poetry," it primarily served as a tool for providing advice about the patent system to readers. In his editor's letter, Porter continues to underscore the journal's commitment to shaping popular discussions of scientific advancements, arguing for the importance of his publication to "the advancement of more extensive intelligence in Arts and Trades in general, but more particularly in the several new, curious and useful arts, which have but recently been discovered and introduced." Citing "communications" from various regions of the country, from the South to the West, Porter defines his audience as "intelligent and liberal workingmen" and positions *Scientific American* as a tool of scientific popularization that will continue to spur innovation in the United States.[19] In his work as editor, Porter emphasized his commitment to making mechanical improvements understandable to readers beyond the rapidly professionalizing scientific community.

From his early manuals on decorative arts to his work on *Scientific American*, Porter's efforts to promote broad knowledge of mechanical and "useful arts" provide significant context for considering the representational strategies he used in crafting his patent model. Unlike the nostalgic murals of New England farms that he painted on the walls of homes and inns, Porter's fire alarm model incorporates his skill at rendering landscapes into an object designed for technological advancement. And while his instructional manual describes the process for creating "rising clouds" against a light blue sky, his patent model instead overlays the landscape with orange flames.[20] Porter also uses his patent model as an opportunity for self-promotion, transforming the wooden box containing the fire alarm's patentable mechanism into a canvas on which to affix his name. Beneath the horizon line of the river passing alongside the house, Porter includes text in the painting's foreground, advertising his own role as an inventor in large capital letters that resemble a title page or newspaper headline. Whether Porter intended for the manufactured product to include these features—particularly the cautionary tale represented in the painting—remains unknown, but the patent model nonetheless clearly blurs the distinction between ornament and function. By illustrating the purpose of his invention, Porter's painting of a house ablaze expands the representational strategies of his model. It not only demonstrates how his invention works but also makes its value more legible to viewers who may be less familiar with the principles of thermodynamics but all too aware of the dangers of fire.

The other forms of patent documentation attached to Porter's fire alarm—the specification and drawing—only underscore the distractions of his patent model. In his "Specification of Letters Patent" and attached drawing (Figure 6.2), Porter emphasizes the internal mechanism, describing the spring that will trigger the alarm's bell. The attached drawing depicts the fire alarm with its door opened to reveal the internal workings, making the external facade invisible—and insignificant—to the viewer's understanding of the workings of the machine, though the specification does mention "a door E, which consists in part of a pane of glass and picture ornament."[21] The specification provides further detail about the process behind Porter's patent: when a metal bar inside the wooden case becomes heated, it will expand and set off a catch-wire that will trigger the movement of the alarm

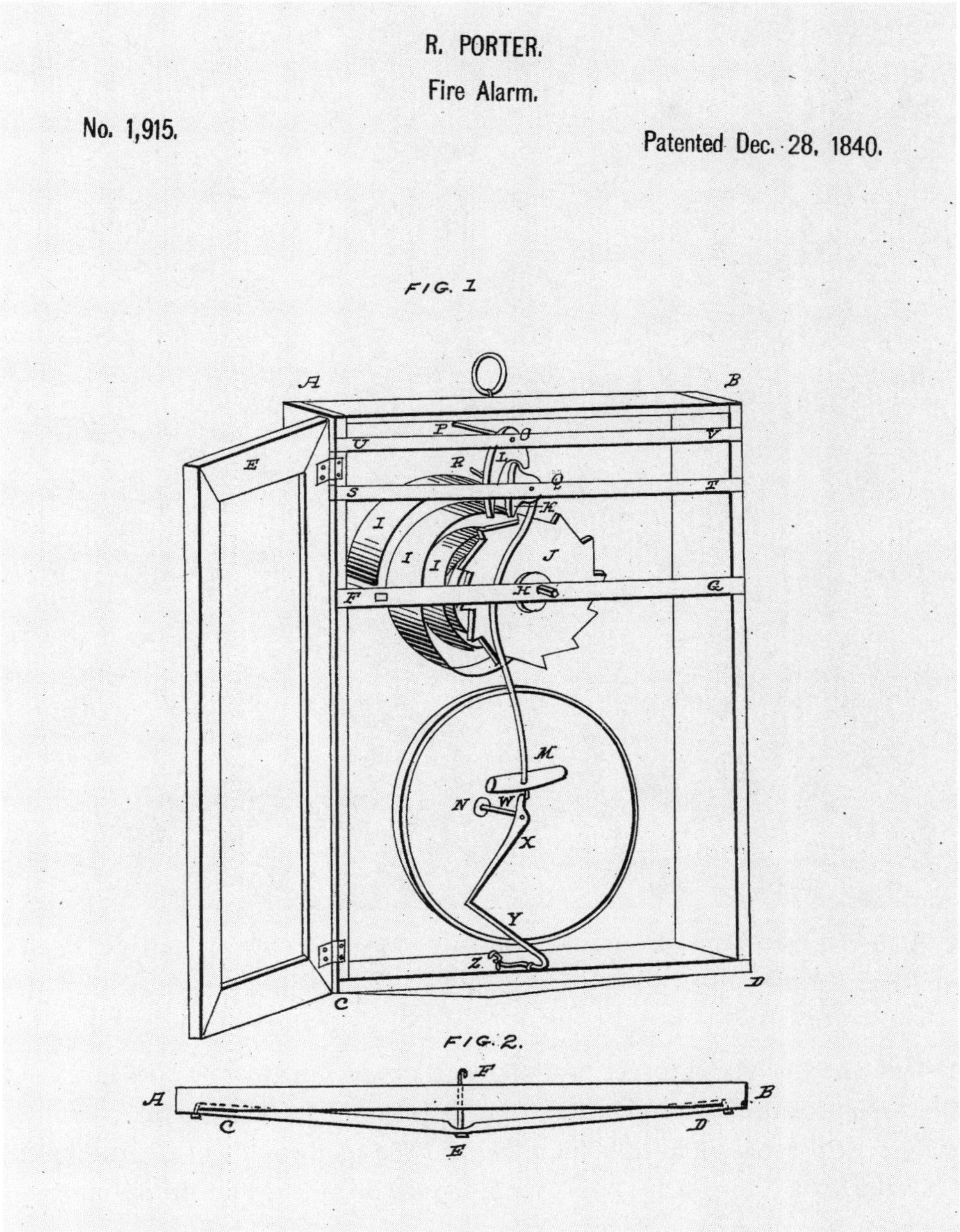

FIGURE 6.2. Patent drawing, fire alarm. Rufus Porter, December 28, 1840. Hagley Museum and Library.

FIGURE 6.3. Patent model, lightning rod. Jacob M. Patterson, July 31, 1860. Hagley Museum and Library.

system. He concludes his specification by acknowledging that he does not claim "the expansion of a bar of metal by heat to disengage or let off an alarm" but rather the precise process rendered in his drawing: "The combination of the metal plate C, D, the hook F, the catch Y, the lever X and the hammer M for the purpose and in the manner described." By focusing entirely on its internal mechanism, Porter's specification reveals the irrelevance of the fire alarm's decorative exterior to its patentability, but its dramatic rendering of a house ablaze nonetheless argues for the usefulness of his invention.

Like Porter's fire alarm, other patent models reveal how inventors incorporated decorative elements to imply an object's function. On July 31, 1860, for example, Jacob M. Patterson of Woodbury, New Jersey, received a patent for "Improved Construction in Lightning Rods" from the U.S. Patent Office. In his application,

Patterson described the purpose of the invention as an aesthetic improvement that allows lightning rods to be attached to the water spout of a house "to save expense and avoid the disfigurement of the walls by the numerous attachments required in the usual mode of erecting lightning-rods and spouts separate from each other."[22] Patterson's patent model (Figure 6.3) aptly illustrates the simultaneous aesthetic deceptions of both his invention and the process of modeling. At first glance, the model appears to represent an entire house, carefully painted red on all sides and in a brick pattern on the front, with painted shutters, gabled windows, a door, and a sloped roof. The lightning rod is just barely visible along the side of the house, lined up with the rain gutter. The model's hidden feature establishes its aesthetic appeal and therefore achieves the purpose of Patterson's patent, showing how this new mode of attaching the lightning rod to the rain gutter will make it as invisible on a life-sized house as it is on the model. Even as the lightning rod suggests how inventors might obscure the most valuable aspects of their invention in their models, it also captures the combined function and aesthetic improvement of Patterson's invention.

Given that most patents were issued for small improvements to existing machines, models like Porter's fire alarm and Patterson's lightning rod reveal the challenges inventors could face in making such minor changes seem exciting to broad audiences. Instead, these examples show how inventors might use patent models to gesture to familiar contexts—the threat of fire, the careful maintenance of a home's exterior—to situate their invention in ways that a lay audience could easily understand. Rather than crafting models that would signal only the technical improvements described in a specification or drawing, inventors chose to blend functionality and aesthetic appeal to make their work legible to as many viewers as possible. Specifications and drawings had to describe a three-dimensional mechanical process in two-dimensional form; however, because patent models were designed to demonstrate how a mechanism worked, inventors could embellish these objects with extraneous information, context, and decoration, as long as they still worked. By pushing the boundaries between relevant information and appealing features, patent models reveal the playful strategies inventors used to signal the value and ultimate marketability of their inventions.

MODELING REFINEMENT

As examples like Porter's fire alarm and Patterson's lightning rod demonstrate, the decorations and contexts added to patent models could both attract and distract viewers. Shifting the focal point of a patent model away from a hidden mechanism might clarify the overall significance of an invention, but it nonetheless raises questions about what kinds of information were most relevant to include in these models.

FIGURE 6.4. Patent model, rocking chair. Charles L. Bauder, April 10, 1842. Hagley Museum and Library.

Even as publications like *Scientific American* reveal the gradual popularization of ideas about technology and invention in print culture, the material construction of patent models shows how inventors similarly sought to reach broader audiences by including advertisements and ornaments that would suggest their later manufacture and consumer use. In her study of nineteenth-century middle-class homes, Katherine Grier has examined advertisements, trade catalogs, and periodicals to show how improvements to the springs, upholstery, and construction of parlor furniture informed broader notions about comfort and gentility.[23] Her work suggests the continuities between mechanical improvements and affective ideas of beauty and refinement, and it further opens up possibilities for considering how inventors sought to anticipate the desires of American consumers in crafting their patent models. By examining the decorative elements added to patent models for a rocking chair and stove, we can see how features that were wholly unnecessary to patent applications played a crucial role in appealing to consumer desires for quality and refinement.

Some decorative elements, for example, existed solely to convey the potential comfort that would result from an improved invention. Charles L. Bauder's model of a "rocking chair" (1842) (Figure 6.4), for example, incorporates orange velvet upholstery and carved, polished wooden arms. On this cushioned chair model, both the chair's back and its footrest extend to demonstrate the process of reclining. The lush texture of the velvet on the seat, back, and armrests appeals to tactile senses even as it visually contrasts with the dark polished wood, implicitly suggesting the quality of craftsmanship and beauty of its external appearance. Such features contrast with Bauder's patent drawing and specification, which describe the invention as an improvement to the frame of the chair, particularly "the mode of constructing the hinge which connects the back to the arms" and "the combination of the sliding rest frame and seat and the combination of the foot board and rest frame."[24] These improvements enable the sitter to adjust the height of the back of the chair and extend the footrest through a set of pins and hinges attached to a notched frame, implicitly improving the sitter's comfort. Even so, these features would not be immediately apparent from looking at the model; an examiner might have to adjust the object herself to see how these mechanisms worked.

By rendering the "rocking chair" patent in polished wood and velvet upholstery, Bauder's model not only demonstrates the working mechanism but also implies the affective result of his invention. Unlike Patterson's lightning rod, the purpose of Bauder's patent is not to increase the external aesthetic appeal of a chair; rather, it is an invisible internal mechanism that will improve the chair's functionality and the sitter's experience. In her reading of parlor chairs, Grier has emphasized the challenge of getting comfortable in nineteenth-century chairs, which were often designed to force the sitter upright and frequently included tough upholstery. Tracing the invention of spring-seat upholstery, she argues that it enabled greater elasticity and bodily comfort.[25] Through soft upholstery and the padded arms of the chair, Bauder's model suggests such comfort, inviting viewers to imagine the pleasure they might feel from resting in such a chair. Moreover, his choice of polished wood and velvet implicitly suggests the high quality and potential expense of such a chair. However advanced its internal mechanism, the inventor suggests, the external appearance of the rocking chair will make it at home in any genteel parlor.

Bauder's decision to enhance the aesthetic appeal of the rocking chair signals his own participation in the marketing process, as the patent model becomes a miniature, not of a patentable improvement, but of the larger machine of which it is a component part. Susan Stewart has emphasized the relationship between the miniature and interiority, arguing that "the miniature becomes a stage on which we project, by means of association or intertextuality, a deliberately framed series of *actions*."[26] Stewart connects the miniature to childhood and nostalgia for an implicitly preindustrial past in which "the miniature object represents an antithetical mode of production: production by the hand, a production that is unique and authentic."[27] Patent models, however, occupy the opposite relationship to temporality, as they represent the anticipation of future manufacture and reproduction. As scale models of future inventions, the patent models indicate a prospective, rather than retrospective, process of making. By crafting aesthetically appealing objects that resemble the familiar trappings of a middle-class home, inventors imagined the eventual marketing and purchase of their inventions.

While Bauder's rocking chair signals affective qualities of comfort and gentility, the ornamental features attached to other patent models both attract and distract viewers from their patentable features. In the case of a stove model by Francis L. Hedenberg (Figure 6.5), for example, the maker applied gilt ornaments that contrast with the black stovepipe and transform an otherwise plain exterior into an aesthetically appealing object. An inventor from New York, Hedenberg was issued a patent on May 7, 1845, for "certain new and useful Improvements in Cylinder-Stoves for Heating Apartments," specifying "the particular manner . . . in which I arrange and combine the flue and air-heating spaces, and the pedestal of my stove, the hot air space being between the ascending and descending draft; the descending draft spreading around the base of the stove."[28] Like Bauder's rocking chair and Porter's fire alarm, Hedenberg received a patent for improvements to a machine's internal mechanism, but his patent model nonetheless features an ornamental exterior.

While Hedenberg's inclusion of such ornaments may seem incidental to the purpose of the patent model, they signal his investment in contemporary trends for household machines, particularly parlor and cooking stoves, which often featured such decorative finishes. Trade catalogs most often emphasized the quality functioning of a stove. But some also occasionally highlighted the ornate patterns that decorated their exteriors, linking its external appearance directly to the quality of its more functional features.[29] One catalog, for example, described the "many new Patterns" that had recently been added as a marker of "the high standard of excellence" of the company's production.[30] Other trade catalogs, such as Shear Packard & Co. in Albany, New York, interweave praise of its external and internal components, describing the "beautiful Gas-burning arrangements" in aesthetic terms as "all that can be desired in a First Class Cooking Stove."[31] The catalog similarly links external "finish" to the quality of the stove's components, describing the castings as "very heavy, very smooth and finely finished" in retaining heat.[32] The exterior ornaments of a stove, including ornamental carvings, played a key role in advertising, allowing manufacturers to signal the high quality of the interior workings.

FIGURE 6.5. Patent model, stove. Francis L. Hedenberg, May 17, 1845. Hagley Museum and Library.

While trade catalogs demonstrate the broader efforts of manufacturers to create "very smooth and finely finished" products, the imperfections on Hedenberg's model also highlight the challenges of constructing models faced by the many inventors who sought to realize their ambitions outside of professional and institutional settings. Each individual decorative element has been applied slightly askew, and they appear to be repurposed gilt decorations, perhaps taken from the keyholes of a dresser or box. These decorations suggest the work of an individual or even amateur maker, rather than a more sophisticated craftsman. Even as they distract from the purpose of representing the patent model's internal mechanisms, they also create an external finish that fails to measure up to the quality of later, finished machines. Hedenberg's ornaments remain aspirational, conveying the inventor's desire to keep up with market trends as well as his limited resources. His attempt to convey the polished quality of his invention might fall short, yet it nonetheless reveals how he sought to participate in modeling strategies that emphasized aesthetic appeal as well as mechanical improvements.

To illustrate a mechanism, it was not necessary for an inventor to include his name and address in splashy, bright-colored lettering along the side of the model; to paint decorative curlicues along the edges; or to include lavish materials such as velvet, polished wood, or brass. Nonetheless, these seemingly extraneous features reveal how inventors used these objects as opportunities not only to demonstrate the workings of a seemingly small improvement but also to showcase the potential for new technology to provide comfort and beauty. The features most irrelevant to a successful patent application were perhaps the most significant to marketing an invention and explaining its value to potential manufacturers and consumers. In this way, patent models reveal how inventors were thinking prospectively, seeking to link minor improvements to affective notions of comfort, gentility, and economic success. These objects not only demonstrate the mechanical processes for which inventors received patents but also emphasize the link between an idea and its imagined realization.

Nonetheless, patent models do raise questions about what features are most significant: the hidden mechanism for which an inventor sought a patent or the eye-catching decoration on its exterior. What, exactly, were patent models intended to model: a newly conceived, patentable process or its imagined form as a consumer product? Ultimately, the three-dimensional format of these objects enabled them to achieve both purposes, allowing inventors to account for multiple kinds of audiences, from patent examiners to potential manufacturers, consumers, and users. By investing these objects with numerous kinds of information and decoration, inventors could allow for multivalent interpretations of their functional and aesthetic appeal and invite viewers to imagine the results of their improved inventions.

NOTES

1 Charles Peirce, "What Is a Sign?," in *The Essential Peirce: Selected Philosophical Writings (1893–1913)* (Bloomington: Indiana University Press, 1998), 5.

2 See Soraya de Chadarevian and Nick Hopwood, eds., *Models: The Third Dimension of Science* (Stanford, Calif.: Stanford University Press, 2004). On the visual properties of models and their role in display, see Ludmilla Jordanova, "Material Models as Visual Culture," in Chadarevian and Hopwood, 443–52. On models, authenticity, and artificial constructions, see Lukas Rieppel, "Bringing Dinosaurs

Back to Life: Exhibiting Prehistory at the American Museum of Natural History," *Isis* 103, no. 3 (2012): 460–90, and Michael Rossi, "Fabricating Authenticity: Modeling a Whale at the American Museum of Natural History, 1906–1974," *Isis* 101, no. 2 (2010): 338–61.

3 On material culture and the history of science and technology, see Lorraine Daston, ed., *Things That Talk: Object Lessons from Art and Science* (New York: Zone Books, 2004). On production, consumption, and the domestic sphere, see Laurel Thatcher Ulrich, *The Age of Homespun: Objects and Stories in the Creation of an American Myth* (New York: Alfred A. Knopf, 2001); Richard Bushman, *The Refinement of America: Persons, Houses, Cities* (New York: Vintage Books, 1992); and Ann Smart Martin, *Buying into the World of Goods: Early Consumers in Backcountry Virginia* (Baltimore: Johns Hopkins University Press, 2008).

4 On the other forms of representation—the specification and drawing—that inventors included with patent applications, see Kara W. Swanson, "Authoring an Invention: Patent Production in the Nineteenth-Century United States," in *Making and Unmaking Intellectual Property: Creative Production in Legal and Cultural Perspective,* ed. Mario Biagioli, Peter Jaszi, and Martha Woodmansee, 41–54 (Chicago: University of Chicago Press, 2011), and on the patent drawing, in the same volume, see William Rankin, "The 'Person Skilled in the Art' Is Really Quite Conventional: U.S. Patent Drawings and the Persona of the Inventor, 1870–2005," 55–78.

5 Johanna Drucker, *SpecLab: Digital Aesthetics and Projects in Speculative Computing* (Chicago: University of Chicago Press, 2009), 6. On interfaces, see Alexander Galloway, *The Interface Effect* (Malden, Mass.: Polity Press, 2012).

6 Patent Act of 1836, sec. 6.

7 Patent Act of 1836, sec. 6.

8 Patent Act of 1836, sec. 6.

9 Barbara Suit Janssen, *Patent Models Index: Guide to the Collections of the National Museum of American History, Smithsonian Institution,* Smithsonian Contributions to History and Technology 54 (Washington, D.C.: Smithsonian Institution Scholarly Press, 2010), viii.

10 Janssen, v.

11 Alain Pottage and Brad Sherman, *Figures of Invention: A History of Modern Patent Law* (New York: Oxford University Press, 2010), 13.

12 George Ticknor Curtis, *A Treatise on the Law of Patents for Useful Inventions in the United States of America* (Boston: Charles C. Little and James Brown, 1849), 131.

13 See Kara Swanson, "Rubbing Elbows and Blowing Smoke: Gender, Class, and Science in the Nineteenth-Century Patent Office," *Isis* 108, no. 1 (2017): 40–61.

14 On the Patent Office gallery, see Reed Gochberg, "Novel Inventions: Emerson, Whitman, and the Patent Office Gallery," *J19: The Journal of Nineteenth-Century*

Americanists 5, no. 1 (2017): 107–28, and Charles F. Robinson, *Temple of Invention: History of a National Landmark* (London: Scala Arts, 2006).

15 See Kara Swanson, "The Emergence of the Professional Patent Practitioner," *Technology and Culture* 50, no. 3 (2009): 519–48. On the close ties between *Scientific American,* Munn & Co., and the U.S. patent system, see Robert Post, *Physics, Patents, and Politics: A Biography of Charles Grafton Page* (New York: Science History, 1976).

16 Rufus Porter, *A Select Collection of Valuable and Curious Arts, and Interesting Experiments, Which Are Well Explained and Warranted Genuine, and May Be Performed Easily, Safely, and at Little Expense,* 5th ed. (Concord, Mass.: Rufus Porter; William Brown, Printer, 1826), iii.

17 On Porter's biography, see Jean Lipman, *Rufus Porter Rediscovered: Artist, Inventor, Journalist, 1792–1884* (New York: Clarkson N. Potter, 1980). Porter's "airship" also coincided with a British inventor's application for a patent for a flying machine in 1843; see Giorgio Riello, "Things That Shape History: Material Culture and Historical Narratives," in *History and Material Culture: A Student's Guide to Approaching Alternative Sources,* ed. Karen Harvey, 24–46 (New York: Routledge, 2009).

18 Rufus Porter, "Announcement," *Scientific American,* August 28, 1845.

19 Rufus Porter, "To the American Public," *Scientific American,* August 28, 1845.

20 Porter, *Valuable and Curious Arts,* 41.

21 Rufus Porter, "Fire-Alarm," U.S. Patent 1,915, December 28, 1840.

22 J. M. Patterson, "Improved Construction of Lightning Rods," U.S. Patent 29,398, issued July 31, 1860.

23 Katherine C. Grier, *Culture and Comfort: Parlor Making and Middle-Class Identity, 1850–1930* (Washington, D.C.: Smithsonian Institution Press, 1988).

24 Charles L. Bauder, "Rocking-Chair," U.S. Patent 2,543, April 10, 1842.

25 Grier, *Culture and Comfort,* 137.

26 Susan Stewart, *On Longing: Narratives of the Miniature, the Gigantic, the Souvenir, the Collection* (Durham, N.C.: Duke University Press, 1993), 54.

27 Stewart, 68.

28 Francis L. Hedenberg, "Stove," U.S. Patent 4,032, May 7, 1845.

29 See *"America." Patent Double-Acting Flue. First-Class Cooking Stove, for Anthracite or Bituminous Coal, or Wood. Patented May 13, 1868* (Albany, N.Y.: William Doyle, [1868]), and *Catalogue of Stoves and Hollow Ware, Manufactured by John F. Rathbone; Office and Sample Rooms, Nos. 9 and 11 Green Street, Foundry, North Ferry Street, Albany, N.Y.* (Albany, N.Y.: Printed by C. Van Benthuysen, 1861).

30 "Circular," in *Catalogue of Stoves and Hollow Ware.*

31 *The American Hot Air, Gas Burning Cooking Stove, Manufactured by Shear Packard & Co., Albany, N.Y.* (Albany, N.Y.: Printed by C. Van Benthuysen, [1863]), 12.

32 *American Hot Air, Gas Burning Cooking Stove,* 22.

7

Bodies Made of Numbers, Numbers Made of Bodies

CATHERINE NEWMAN HOWE

The plaster sculpture of the "average" man leans back casually on his left leg with one hand behind his back, arm akimbo, the other resting at his side in a calm, assured and naturally confident stance. He is slender, clearly muscular but not muscle bound, well proportioned, and carefully coifed. His modesty is preserved not by the traditional fig leaf but by a more ornamental, Beaux-Arts frond that sweeps gracefully across his hips (Figure 7.1). The sculpture of the "average" woman, too, is slender. Soft curls rest on her brow, while the rest of her hair is pulled back into a loose chignon. Leafless, her modesty is preserved by the conventional classical absence (Figure 7.2). While the young man seems to be posing for a full-length portrait, she is caught in motion. Both sculptures adhere to a kind of vague surface classicism in that they mimic the style of modeling and pose and approximate the surface treatment of classical Greek, Roman, and neoclassical works.[1] Their proportions would seem to be based on familiar canons, idealized mathematical ratios codified by the likes of Pliny and Vitruvius, but they are not. They could be mistaken for sculptural studies or plaster casts, but they are original, and they are complete. Even to the trained eye, it is not apparent that their figures are modeled on anthropometric statistics or that their faces are based on composite photographs. In anticipation of the sculptures' unveiling, one newspaper headline declared "The Perfect Physical Man: Two Boston Sculptors at Work on the Ideal Forms of Modern Life—College Students Furnish the Measurements—the Scientific Mean Taken in All Instances." The enthusiastic author of the article notes that the purpose of the sculptures was to "secure a model of the most perfect forms ideally of modern life."[2] This, however, was decidedly not the case. On the contrary, the sculptures were intended precisely *not* to depict "the most perfect forms ideally." They were meant to model scientifically and artistically what *was,* not what *could* be.

These two sculptures—scientific models playing the game of art, art objects playing the game of science—were commissioned from the Boston-based artists

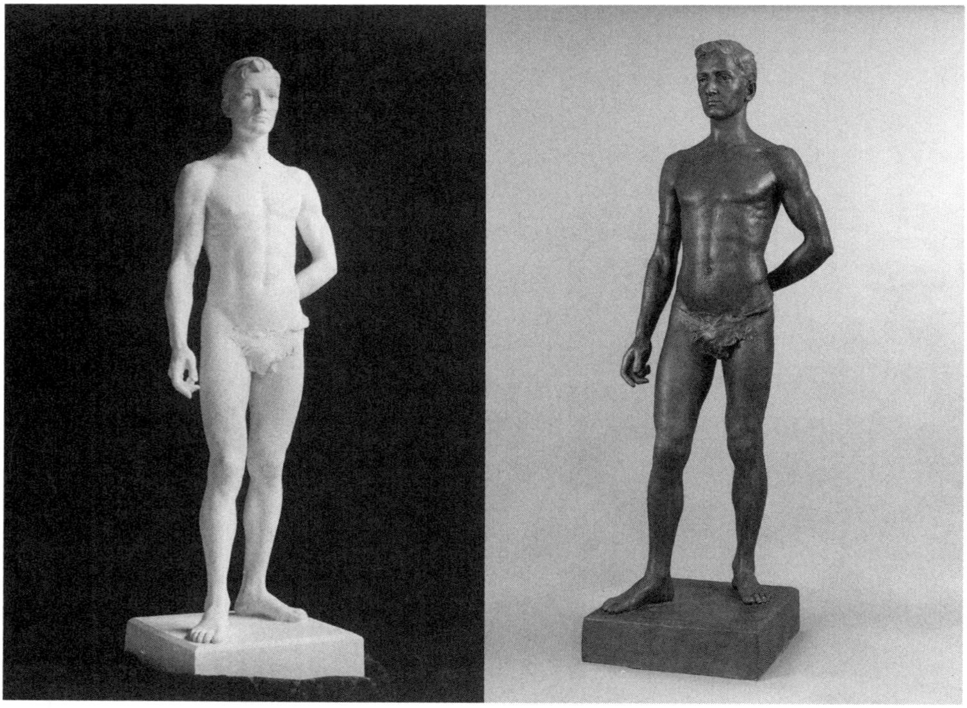

FIGURE 7.1. Henry Hudson Kitson and Theo Alice Ruggles Kitson, *Composite Figure of Typical American Man* (1892). Plaster (Boston Public Library). Painted bronze. Gift of Dr. Dudley A. Sargent, 1920. Copyright President and Fellows of Harvard College, Peabody Museum of Archaeology and Ethnology, PM 20-3-10/59929.0.

Henry Hudson Kitson and Theo Alice Ruggles in 1892 by Dudley Allen Sargent, head of Harvard University's Hemenway Gymnasium and director of Harvard's fledgling physical fitness program, which emphasized a regimented approach to exercise and a calculated view of the body. For Sargent, fitness meant more than strength for labor or sport; rather, fitness was a science.[3] Based on the anthropometric measurements of thousands of college students from Harvard University, the Harvard "Annex" (later to become Radcliffe), and several other elite universities and colleges, the sculptures were the physical embodiment of Sargent's carefully calculated average American standard and the outcome of contrived, three-dimensional visions of numeric models.[4] Once complete, the sculptures were sent to Chicago to be part of an ambitious display at the 1893 Columbian Exposition, a world fair conceived as an opportunity to confirm Chicago's status as a major city and as a chance for the country as a whole to mark its emergence as a global leader in science, technology, and the arts. At the Exposition, the sculptures were strategically placed models intended to act as dual signposts, one for orientation linking the idealized sculpted bodies that were brazenly deployed in the glistening White City,

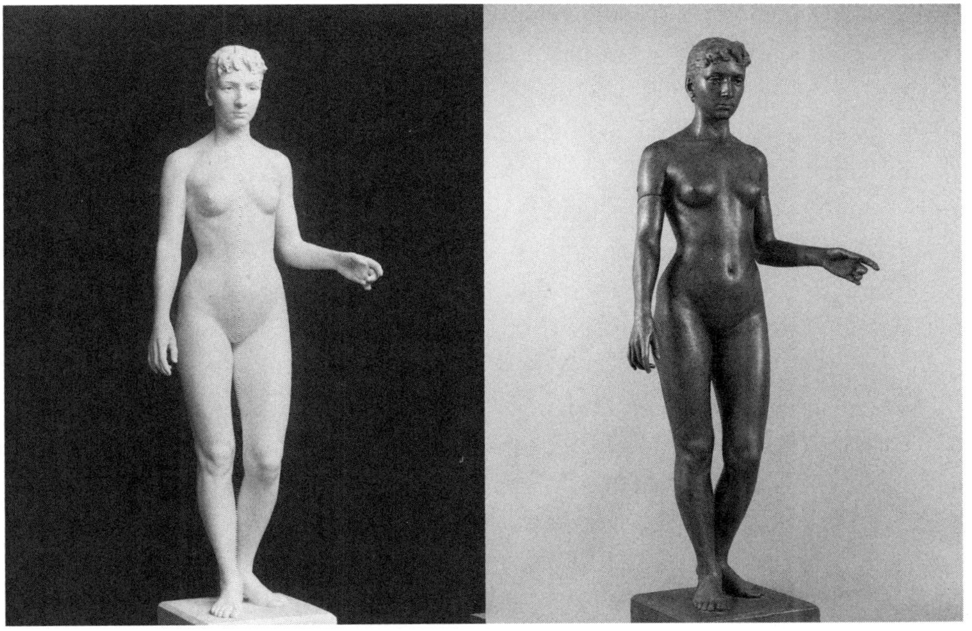

FIGURE 7.2. Henry Hudson Kitson and Theo Alice Ruggles Kitson, Composite Figure of Typical American Woman (1892). Plaster (Boston Public Library). Painted bronze. Gift of Dr. Dudley A. Sargent, 1920. Copyright President and Fellows of Harvard College, Peabody Museum of Archaeology and Ethnology, PM 20-3-10/59930.0.

the other for navigating the exoticized—and very real—bodies on display at the Exposition's infamous Midway Plaisance.

The White City, a grandiose cluster of whitewashed neoclassical buildings and in many ways the centerpiece of the Exhibition, was an extravagant celebration of Beaux-Arts style, designed to visually link the United States with the powerful empires of antiquity and to showcase the best that American civilization had to offer. The sculptural program of the White City, under the guidance of Augustus Saint-Gaudens, was ambitious and vast and gathered what Saint-Gaudens considered to be the greatest assembly of artists since the Renaissance. They included Daniel Chester French, whose colossal *Republic* dominated the main axis of the fair; Frederick MacMonnies, who created the *Barge of State* or *Triumph of Columbia*; and Karl Bitter, who created the monumental figural groups for the Administration Building.[5]

The White City stood in stark contrast to the exoticism of the Midway Plaisance, where visitors could visit an expansive installation of simulated native villages where fairgoers could see living and breathing exhibits of the "races of the world." Visitors could be entertained by Arabs, Japanese, West Africans, and Polynesians, among many others, some of whom were presented in "natural" living dioramas. The Midway was organized to create the impression of a "human kaleidoscope," but it nevertheless confirmed one of the basic principles of social Darwinism—that there existed a hierarchy of civilizations and races that moved from the African "savage" to the "enlightened" Anglo-Saxon. The Midway staged and naturalized a teleological view of cultural and racial development. It was where visitors went to see the "uncivilized other" and a place where difference was imagined, celebrated, and exploited. But if visitors wanted to see the familiar, in this case, the "objectively accurate" models of Americans, they were directed to Sargent's statues.[6] These complemented the evolutionary spectacle on display at the Exposition as a whole, but with a twist: Sargent's statues promoted the idea that the physical progress of the individual was explicitly linked to the progress of the nation.

Sargent's decision to hire academically trained artists with established reputations to translate his statistics into sculptural form was motivated by an awareness of the didactic limitations of both tabular and graphic representations of scientific data as well as his understanding of the discursive power of the classical sculptural tradition. If a scientific model is designed to represent visible phenomena that cannot be experienced directly, while an artistic model is designed to represent otherwise elusive aesthetic possibilities, Sargent's sculptures arguably do both. In yoking the supposedly objective power of anthropometric statistical models to the subjective power of artistic ones, Sargent was able to produce a set of statues that effectively translated the results of his scientific labors into an accessible visual model built on a broadly comprehensible classical vocabulary. They demonstrate how the combination of statistical and sculptural modeling can result in a powerful form of material storytelling, complete with all of its connotations of truthful representation despite being embedded in creative fabrication.

Hired by Harvard University in 1879 and buoyed by its authority, Sargent made it his lifelong project to inspire people to achieve a certain physical and mental balance that by his lights ought to be the proper sign of health. Like many of his contemporaries, he was concerned about a perceived deterioration of both the male and female American body. The stress of an increasingly industrialized modern life, the devastation of the Civil War, and the spread of social Darwinist thought had formed a discourse of deterioration, a discourse according to his contemporaries that also blamed urbanization and immigration for disrupting firmly entrenched gender roles, family relations, and sex in late nineteenth-century America.[7] Of particular concern was the condition called "neurasthenia," a "nervous disease" identified by the neurologist George M. Beard in 1869 that manifested itself in headaches, depression, and muscular weakness, and which doctors linked directly to an increasingly industrialized modern life affecting men and women particularly in the middle and upper classes.[8]

A diverse number of professionals, such as the physician and educational reformers William Andrus Alcott, the minister Sylvester Graham, and the nutritionist John Harvey Kellogg (Sargent's contemporary), reacted to this epidemic by spearheading broader health reform initiatives that emphasized diet over exercise and that were tied more explicitly to spiritual and moral concerns. Electrotherapy devices, mineral water, new patent medicines, hydropathy, chiropractic, and sanitaria were also popular. By contrast, Sargent was one of the few who suggested that health could be achieved through physical exercise, and his fitness program was developed as the antidote to a perceived national crisis. He was convinced that the most practical way to ward off neurasthenia and similar modern ailments was by balancing body and mind through a rigorous program of physical fitness. The ultimate goal of his program was not athletic success. Rather, it was to encourage the strength and physical symmetry required to support his more comprehensive vision of good health.

At Harvard, Sargent lectured on hygiene, physical development, and applied anatomy. But his greatest interest lay in anthropometry, the systematic collection of measurements of the human body for comparative analysis and a field of study that he considered critical for his work and that he would go on to refine and expand. During his long tenure, Sargent invited all undergraduates to undergo a thorough physical examination, which enabled him not only to record their physical measurements following a standardized protocol but also to document physical changes over time. According to Sargent, 87 percent of the students "avail[ed] themselves of this privilege."[9] The examination included more than forty measurements taken with tape measures and calipers, manometers and spring dynameters, most of which were designed by Sargent and his assistants. Over time, more than eighteen thousand students participated in this study, generating a carefully crafted empirical record consisting of examination cards and managed by a meticulous filing system.[10] Armed with their measurements, students could tailor their fitness regimes using the "Sargent System," a set of exercise machines designed by Sargent to target specific muscle groups with pulley weights that were, Sargent wrote, "adjustable to the strength of the strong and to the weakness of the weak."[11] If the machines were not available, those interested in physical improvement could eventually find guidance in the books Sargent published: *Universal Test for Health, Speed and Endurance* (1902), *Health, Strength and Power* (1904), and *Physical Education* (1906).

While the anthropometric exam guaranteed "accuracy" as long as the examiner followed Sargent's instructions, Sargent himself was concerned that the results were limited in practical use for two reasons. First, Sargent suspected that numerical results alone were not easily digested by the nonexpert, which prompted him to explore representational methods that could model his work in a visually accessible and appealing way. Second, the data were valuable to the examined person only if he could compare the results of his exam to those of his peers. Without a standard, human variation had no meaning. A graph could be descriptive without a central line representing the average, but it could not be prescriptive. Without that line, how could a student know where he stood or what she needed to do to achieve a more balanced, symmetrical body? Sargent solved both problems by creating an average standard against which his subjects could be compared and by constructing an easily comprehensible graphic template upon which every individual's data could be mapped against that average. In short, Sargent needed a model to create his model, a kind of reasoning that was as circular as it was commonsensical.[12]

Like the doctor who needs to establish a baseline pulse before diagnosing an abnormality of the heart, Sargent needed a baseline body. "We have had no end of treatises on the sports . . . that are reputed to give strength and symmetry to the body," he complained in an article dedicated to the subject of the "typical," "but unfortunately the wise and good men of old have left us no standards by which to judge of symmetry or strength. The ancient masterpieces are models of symmetry and beauty, but they were made largely from ideal standards, certainly not from ideal measurements."[13] Pitting inductive ideals against deductive practices, Sargent used statements like this to distinguish between images of the *ideal body* derived from artistic canons and images of the *ideally measured body* derived from science. If the former were abstracted from theory, he argued, the latter were immanent reality.

When Sargent finally compiled the data points collected from his student specimens and plotted them out on a graph, the results did not disappoint.[14] The distribution of variants did indeed create a normal curve, a symmetrical bell shape formed by the equal and continuous distribution of data around a mean. For Sargent, it was as if "natural law" had been made empirically visible, and it meant that he could confidently establish the average standard upon which the success of his project depended.[15] The chart that he ultimately developed was bifurcated by a vertical line representing the average (Figure 7.3). The line representing that individual's proportions would parallel the vertical line of the standard. As Sargent explained, "the straight line is the physical sign of health and longevity, of perfect structure and harmony of function, and a symmetrical development of the whole body."[16] The average, therefore, could be seen as a kind of ideal model that, paradoxically, left plenty of room for improvement.[17]

According to Sargent, the object of the chart was to give young men and women an incentive to exercise by showing them at a glance where they stood in relation to a "normal standard," and he urged those whom he measured and who adopted his program not to compete with those of superior abilities but rather for each

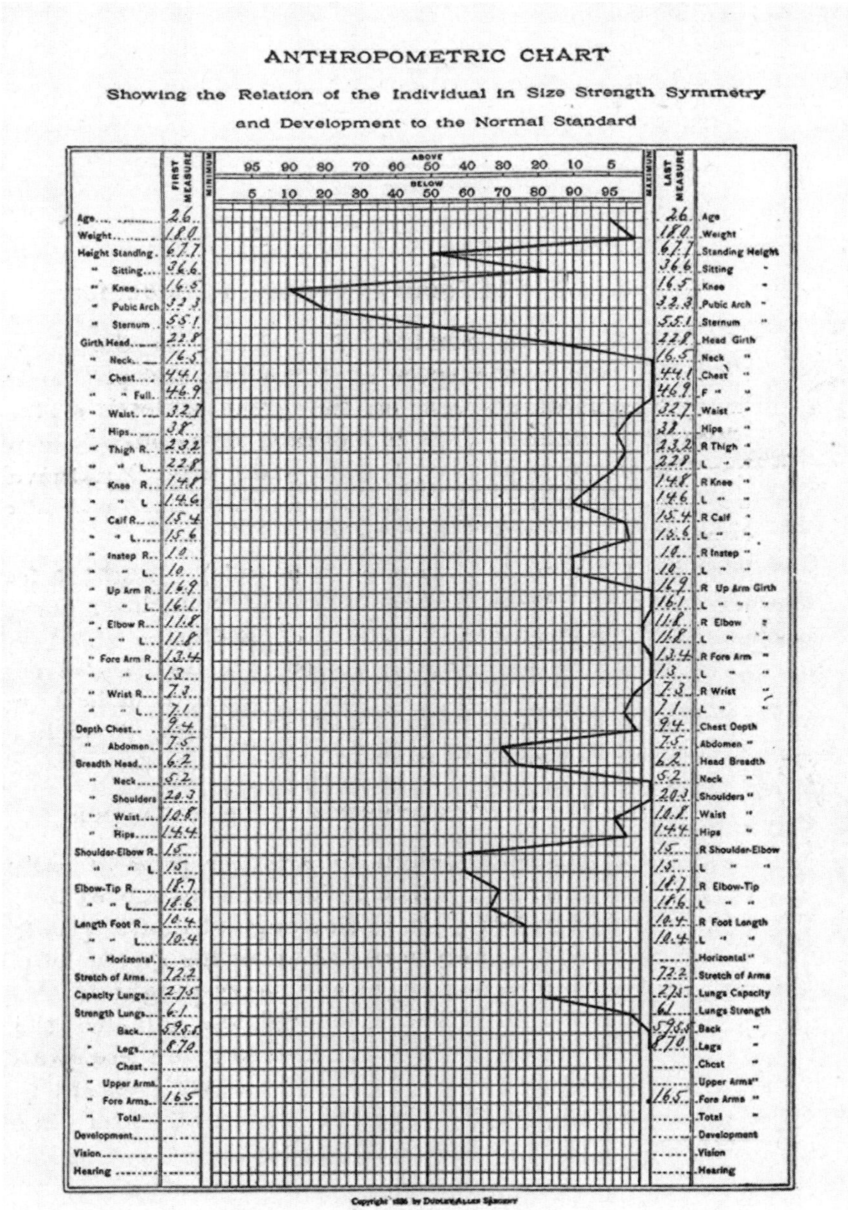

FIGURE 7.3. Dudley A. Sargent's anthropometric chart for Eugen Sandow. Printed in G. Mercer Adam, ed., *Sandow's System of Physical Training* (New York: J. Selvin Tait, 1894), 241.

individual to "compete with his own physical condition" in the hope of achieving a *personal* ideal. Sargent was practical enough to understand that few, if any, would be able to achieve a body that would result in a straight line to the far right of the standard, no matter how dedicated he was to Sargent's carefully designed, mechanized system of exercise. Even the weakest person could reach her own potential if she faithfully adhered to the program. A perfectionist with a pragmatic spirit, Sargent hoped his students would strive for a measurable—and ultimately visible—physical equilibrium, not flawlessness, as the ultimate goal.[18]

Graphs and tables had their own limitations, however. Sargent's chart, while supposedly straightforward and uncomplicated, was still tailored to a specific individual. Finding a representative figure of the ideal in the flesh would have been a public relations coup. That theoretically perfect man appeared when Sargent was asked by Joseph Pulitzer's *New York World* to examine Eugene Sandow, the legendary strongman and entertainer. According to the *World,* Sandow was "not only inspiring because of his enormous strength, but absolutely beautiful as a work of art as well. . . . One look at him is enough to make the average young man thoroughly disgusted with himself, and make him give up his nightly habit of standing in front of the glass in his pajamas and swelling his chest with pride."[19] His examination led Sargent to conclude that "Sandow is the most wonderful specimen of man I have ever seen. . . . He is strong, active and graceful, combining the characteristics of Apollo, Hercules, and the ideal athlete."[20]

Though Sandow's chart was not perfect (his diminutive knee, for example, was well below the average, while all of his other measurements were significantly above), he was certainly the visual approximation of ideal numbers in the flesh. The press saw Sandow simply as "a work of art," frequently comparing him to classical statues. Sandow, who frequently referred to his own study of Greek and Roman sculpture, encouraged such comparisons by regularly harnessing the iconography of antiquity in his visual representation. In meticulously composed publicity photographs, Sandow donned gladiator sandals and fig leaves; powdered his hair and body for a more marble-like, "sculptural" look; used pedestals and classical columns as props; and assumed poses reminiscent of such well-known sculptures as the Farnese Hercules and the Dying Gaul. In 1901, Sandow was even selected to be the model for a sculpture of his own, a cast representing the "perfect type of a European man."[21] One cast of the strongman was bought by the British Museum. A second was displayed in a vaudeville theater in Boston before Sandow personally gave it to none other than Dudley Allen Sargent.[22] Sandow, whether in the flesh, in photographs, or in plaster, was the gold standard. His was the body to which American men were supposed to aspire.

Sargent was as impressed by Sandow as the general public, yet it was the question of what the *average* statistics looked like in three dimensions that intrigued

Sargent most. "If any one man could be found," he wrote, "all of whose measurements corresponded to those on the central line in the table, he would be termed a *mean* or typical man."[23] Though Sargent had found someone who came close to an ideal in Sandow, he had found no one person that even approximated the measurements of his elusive average. As one article about his project noted, "an odd feature has been that scarcely any individual filled the measurements of the type in more than one or two particulars and a great multitude, of course, have not one of the measurements of the typical figure."[24] And yet, to capture the public's imagination, Sargent was convinced that his average standards must have bodies. This required a shift in representational modes, from data points on a chart to three-dimensional form. Unable to find models of the average in the flesh, Sargent turned to the closest alternative he could find: sculpture.

Through sculpture, Sargent found a way to make statistical reality visible, even if it was a reality that could not be found in nature. The sculptures embodied a mimetic paradox: "realistic" representations of bodies that did not exist. Like many conceptual models, they embodied a *theoretical* vision of the average, derived from concrete points of data. In some ways, this was no different from the idealized figures based on traditional artistic canons, conceived of as composites of idealized particulars, yet Sargent's figures' figures were restricted by statistics rather than artistic regimes. At the same time, sculpture allowed him to fuse his statistical data with a classical artistic lexicon, one that was not only comprehensible to fairgoers but the dominant decorative motif of the fair as a whole. In addition, the very materiality of sculpture was a way to engage an audience in a way that two-dimensional representations could not. If Sargent could not turn his numbers into flesh, he could at least turn them into its visual approximation.

Crucial to producing a model body was the specificity of the medium. Because sculpture had a tangible, tactile allure, Sargent anticipated that visitors to the Columbian Exposition would be tempted to experience the model of the typical American man or woman close up, perhaps grasping the biceps despite the rope barriers that separated them from their plaster representatives. They could potentially envision their *own* bodies as sculpture. But while he wanted to take full advantage of the materiality and viewing experience unique to the sculptural medium, Sargent also saw the sculptures as an important social corrective, an antidote to the kinds of civic sculpture that were usually available to the general public. Sargent worried that Americans were in real danger conflating the real and the ideal. He argued that the "splendid specimens that often appear (undraped) upon the floats of our boat houses, [and] the athletic fields" were distorting the public perception of the typical American body. Sculptures whose proportions were strictly based on statistics, on the other hand, would presumably avoid the pitfalls of heroism and of romanticism and could provide the general public with a much-needed visual corrective.[25] Nowhere was that needed more than at the Columbian Exposition, where the ideal sculpted body was an integral part of the fair's spectacle and broadly deployed to assert racialized nationalistic ideals.[26]

While Sargent might not have conceived of his project in overtly racist or classist terms, his average standards, meant to apply to all Americans, were still based upon data derived from a small, elite, white, well-nourished (though underexercised, according to Sargent) segment of society. Although he encouraged improvement for all, his system plotted the measurements of all subjects against standards derived primarily from the white and well-to-do, those who would have been embraced by advocates of eugenic regimes. The sculptures thus did very little if anything to challenge the underlying assumptions of existing racial "science." Sargent saw his sculptures as statistics transformed into art, not necessarily as reifications of race, yet the racial implications of his statistical sculptures became inevitable once put on display at the Exposition.

It is paradoxical, given his concerns, that Sargent would turn to two artists, Henry Hudson Kitson and Theo Alice Ruggles, who were particularly familiar with the conventions of classical monumental sculpture and unquestionably guilty of creating the heroic types that Sargent found so problematic.[27] Kitson, who earned his artistic credentials studying sculpture at the École des Beaux-Arts in Paris before opening his own studio in Boston, was considered to be one of the best creators of public statuary in New England by the peak of his career.[28] Today, he is best known for his *Minuteman* (1900) statue in Lexington, Massachusetts, and the *William Conant* (1905) statue in Salem, Massachusetts. At the time of the commission, Ruggles was Kitson's most promising art pupil (and future wife), a rising star in the field of monumental statuary. Ruggles, too, built her artistic reputation on war memorials, including *The Volunteer* (1902) to honor those who fought in the Civil War and *The Hiker* (1906) to commemorate American soldiers who fought in the Spanish–American War, the Boxer Rebellion, and the Philippine–American War.[29] By the time Sargent approached Kitson and Ruggles in 1892, he could do so with confidence. Kitson had secured his reputation as a gifted monumental sculptor, and Ruggles was well on her way to doing the same.

Because of its unusual limitations, however, Sargent's commission was not one that many fine artists would have considered. Most artists had little use for the "scientific" reconstruction of a numerically average American body when they could rely on established artistic canons. The classical canon, traced back to the Roman artist Polyclitus and which remains with us to this day, understands the sculpted body as a mathematical equation of abstract proportions. The sculpted ideal is reflective of perfect ratios—each body part in harmonious proportion with every other—rather than a "neutral" mathematic compilation of multiple data points. Sculptures that adhered to the canon, as Michael Squire notes, embodied a "materialized ideal of immaterial beauty... thought to hold universally."[30] The numerical *average,* on the other hand, failed to capture the imagination of artists, even those who worked strictly within the realist tradition, for to depict a figure that represented a statistical average was to violate the aesthetic and ethical norms of "fine art" meant to elevate and inspire rather than simply to record or to document, and of the artist whose job it was to give imagined ideas concrete form.[31] This is an interesting paradox, of course, seeing as the reification of an average, making an abstract concept material, must *always* be an act of imagination.

Kitson and Ruggles also had to work without the benefit and inspiration of live models, a standard practice for trained sculptors, though it wasn't for lack of trying. The artists were greatly frustrated by "the impossibility of finding a model who could come anywhere near the proportions of the figure ordered."[32] Their frustration, of course, only served to confirm Sargent's reasons for the commission. By accepting this commission, however, Kitson and Ruggles were committing themselves to a project that minimized artistic license and intervention. Kitson and Ruggles were constrained by numbers, but giving compelling form to those numbers nevertheless required creativity. Capturing the public imagination required imagination, particularly when it came to the sculptures' faces.

Sargent did not collect anthropometric data relating to facial features, a decision that set him apart from a long physiognomic tradition. The idea of physiognomy, made popular by Johann Caspar Lavatar in the 1770s, was an anthology of facial types that were linked to specific character traits. According to Lavatar's system, the weight of one's brow or the breadth of one's nose could be read like a text to reveal moral rectitude or depravity. Phrenology, which enjoyed great popularity through the nineteenth century, claimed to be able to do the same through the study of the contours of the skull. Physiognomists and phrenologists, professional and amateur alike, were therefore more than keen observers of physical traits; rather, they were moral magistrates.[33] Yet Sargent's choice to eschew the topography of the face made practical sense. After all, his focus was how physical fitness could improve the body, and while the circumference of a bicep could be shaped with pulleys and weights, a prominent nose was forever formed. More important, Sargent was interested in recording physical measurements, not in ascribing moral characteristics to them, at least not consciously, and so sought an alternative, nonjudgmental approach. Basing the faces on composite photographs was a way to maintain the objectivity of the project as a whole. The faces of the sculptures, like their figures, had to be grounded in something objectively *real,* not on moral considerations, not on the aquiline features of classical sculpture or the fancy of the artists' imaginations.

Seeking to avoid artistic license or convention, Sargent decided to model the faces of the sculptures on composite photographs. A contemporary news article about the origins of the sculptures indicates that the faces of the sculptures were based on composites of Harvard and Annex students created by the photographer John L. Lovell, which were then "reinforced" by composite photographs of men at Amherst, Bowdoin, "and other colleges" and of women from "similar, elite female colleges" into "co-composites."[34] While neither Sargent nor Kitson and Ruggles specified the exact "co-composite" photographs that guided them, evidence strongly suggests that they were created from Lovell's work for the Boston photography firm of Allen and Rowell (Figure 7.4). The male co-composite was of 449 men from Harvard, Amherst, Bowdoin, Johns Hopkins, Sheffield Scientific, and Williams College, and the female co-composite was of 287 women from Harvard "Annex," Smith, Wellesley, Wells, and Mount Holyoke.[35] In using these photographs to guide the facial features of the statues, Sargent hoped to create the illusion of quantifiable statistical evidence produced by a seemingly value-free technology.

Yet composite photography was anything but value-free. Its invention is generally attributed to Frances Galton, the British statistician and the founder of eugenics, who presumed that eugenic perfection and deficiency were legible in the body itself and for whom the *visual* representation of the average was a primary concern. Composite photography was for Galton a reliable method for determining whether physiognomy was, as he put it, "an index to the mind."[36] Composites, Galton argued, meant that, for the first time, scientists could make "valid" predictions about different races and classes of people using an authoritative visual index. In short, he considered composites to be the photographic equivalent of large statistical tables "whose totals, divided by the number of cases and entered in the bottom line, are the averages. They are the *real generalizations,* because they include the whole of material under consideration."[37]

Galton further argued that for a composite to be persuasive, it was paramount that the combined images be similar in size and "attitude," that the eyes of each of the photographic subjects be aligned, and that each component photograph be exposed to the photographic plate for the same length of time. The darkest and sharpest outlines in the resulting image would then be those "common to the largest number of components, [whereas] purely individual peculiarities would leave little or no visible trace."[38] When done well, claimed Galton, composites have a "surprising air of reality. Nobody who glanced at one of them for the first time would doubt its being the likeness of a living person."[39] And indeed, amusing stories of such blunders were occasionally recounted in popular magazines. For example, *Outing* magazine (a sportsmen's journal) related the story of a young man who was particularly taken by the portrait of a young woman whose expression "denoted a

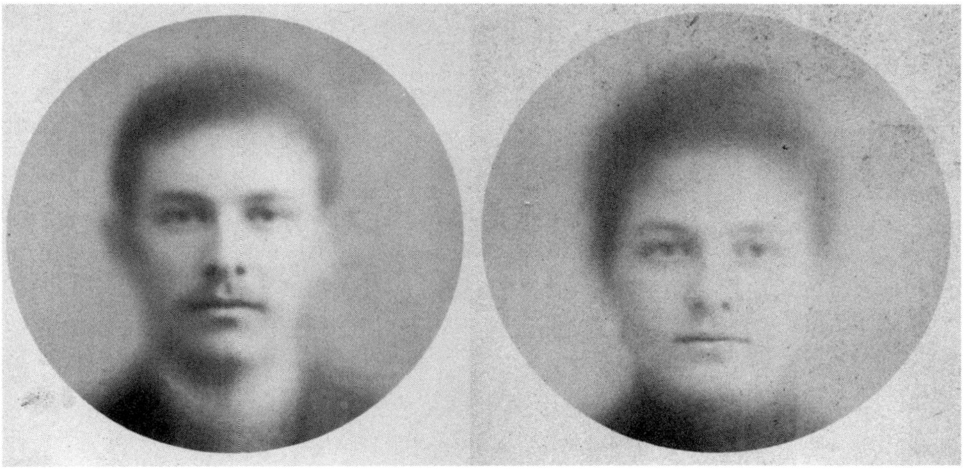

FIGURE 7.4. "Co-composite of Amherst, Bowdoin, Cornell, Harvard, Johns Hopkins, Sheffield Scientific, and Williams composites. A group of 449" and "Co-composite of the 'Annex,' Mount Holyoke, Smith, Wellesley, Wells, and Vassar composites. A group of 287." The quality of the images reflects the originals. Courtesy of Smith College Special Collections.

strong will, yet a gentle disposition." When he inquired as to her identity, he was told, "The graduating class of Smith College." Alas, the young man "awoke regretfully to the fact that there was in reality no such lady as the one whose face had so strongly impressed him, or, rather, that there were forty-nine of her!"[40] A truly winning composite would play the same trick and therefore had to efface the means of its own production as much as possible.

As Alan Sekula has shown in his seminal essay "The Body and the Archive," composite photographs were an attempt to embed an archive into a single image and, by extension, a means to naturalize contemporary racial typologies.[41] Many elite colleges and universities sold cabinet-card prints of composites of individual class years to students as novelty items and sentimental keepsakes. Just because they were not created with an explicit intent to visualize race, however, does not mean that they did not contribute to and perpetuate a problematic taxonomy of the young, elite, white northeasterner. The art historian Kris Belden-Adams has explained how composites of college classes "offered a literal, as well as metaphorical, symbol of cohesion" among the upper-class elite that "subtly underscore[d] the importance of conforming and upholding the late nineteenth century social order." They were images of the "strong-willed" yet "gentle" women of Smith, the Annex, and similar colleges who were being educated to be compatible spouses for their elite male counterparts. They were visual representations of the kind of positive eugenics advocated for by Galton, condensed apparitions of affluent, white heteronormativity.[42]

The search for an accurate way to represent a type was by no means new, but as Galton pointed out, "the imaginative power of even the highest artists is far from precise, and is so apt to be biased by special cases that may have struck their fancies, that no two artists agree in any of their typical forms."[43] An artist's rendition of any general type was therefore suspect. Galton maintained that composite photography could eliminate this source of subjectivity, if not entirely, then at least to a greater degree, relative to the number of component photographs used to compose the final image. While Galton praised composite photographs that were aligned so seamlessly as to trick the eye, he also believed, incongruously, that the unclear contours that gave most composites their haunting quality were not flaws. They might compromise a composite's mimetic veracity, but they also revealed an underlying statistical truth. Thus the blur was of concrete value because it showed the distribution of the component faces around the mean. It was the visual expression of the sloping sides of the bell curve, a form of pictorial statistics. Just like the three-dimensional bodies of Sargent's statues, composite photographs, though entirely artificial, were two-dimensional models promising a version of invisible reality. Sargent's decision to take advantage of the indexical properties of photography, as well as the composite's theoretical ability to translate the credibility of numbers into pictorial form, enhanced the scientific aura of the sculptures.

Practically speaking, of course, sculpting a visage from a composite must have presented a particular challenge for the artists. The blurred outline of a composite photograph, meant to be read as the marker of statistical objectivity, is by necessity lost in translation when transformed into sculpture. Kitson and Ruggles, however, had to draw the lines somewhere. They had to give contour to the blur, to create a border for the statistic and artistic object in space. Hair had to be fashionably coifed. Ears needed to take concrete form. Facial contours had to be clearly legible. This required artistic judgment. Moreover, judging by photographs of Ruggles, it was likely that her own visage and curling locks inspired the female figure's face. Like countless artists before and since, Ruggles might well have used the nearest person at hand, even when nonindividuality was the whole point of the commission. The management of statistical inconsistencies also required artistic judgment. Though the faces of the sculptures were based on composites of a few hundred students, the bodies of the figures were based on the measurements of ten thousand people.[44] As a result, the bodies of the sculptures were effectively "more average" representatives than the faces, a discrepancy that could be concealed, though not resolved, by the sculptors' creative hands.

Despite inconsistency of the data upon which the sculptures were based, the press stressed the sculptures' statistical precision as soon as news of their commission was made public. The *Illustrated American*, a popular national magazine, described the circumstances of the sculptures' creation and noted that Boston was "anxiously awaiting a view" of the "plastic representation of the results of measurements of more than ten thousand persons taken scrupulously by the most approved anthropological methods." It was careful to point out that the sculptures were based on mathematical means. "In this way," it noted, "the extremes have no part in the calculation."[45] One journalist, sent by the *Boston Post* to Kitson's studio to preview the sculptures, was delighted by the pair. He was particularly gratified by their numerical exactitude, and it was their accuracy "to the average type to the infinitesimal fraction of an inch in every portion" that he emphasized in his article.[46] He ignored the sculptures' classical aesthetic entirely, perhaps not surprisingly, however, for when he toured the studio, the sculptures were still works in progress. The male figure was still wrapped in wet linen and "ignominiously stuck full of splints," and the female figure's "toilet" was similarly incomplete.

Another *Illustrated American* feature about Kitson and Ruggles and their collaboration with Sargent is accompanied by a large photograph of the artists in their Boston studio, surrounded by works in progress (Figure 7.5). Kitson directs an assistant from a ladder next to a large plaster statue of an athlete battling an eagle (also destined for the Fine Arts Building at the Columbian Exposition, before being cast in bronze for the Dyer Memorial Fountain in Providence, Rhode Island), while Ruggles, chisel in hand, examines a sculpture of a young boy (destined for the Women's Building at the fair). A live model, the traditional inspiration for the figural arts, stands on a pedestal behind Ruggles. In an interesting juxtaposition, however, Sargent is seated in the foreground of the photograph, overseeing the artists, presumably to make sure that their processes are not *too* creative. Framing the photograph are the two average figures, swathed completely in linen.[47] The editors of *Illustrated American* knew that the mummy-like figures would pique the curiosity of their readers, but showing them in the studio of the artists, surrounded by the kind of grand Beaux-Arts sculpture so popular at the time, had another effect: it strengthened the argument that the sculptures, though based on objective science, were also legitimate art, and it implied that viewers should await their unveiling with the same degree of suspense they would reserve for that of a public monument.[48]

The sculptures were displayed in two forms at the Columbian Exposition: in the flesh (so to speak) and in photographs. Only the photographs were displayed in Harvard University's expansive exhibit in the Liberal Arts Building. The goal of the exhibit was to demonstrate the successful "application of scientific methods" by higher education and included all manner of "statistics, graphic charts, laboratory photographs, specimens and apparatus used in instruction." Even subjects that would seem naturally resistant to such tabular and diagrammatic methods

FIGURE 7.5. Henry Hudson Kitson's studio. Printed in "The Typical Man and Woman of America," *Illustrated American,* June 17, 1893, 708.

of display, for example, the works of the Classical Department, were reduced to numbers.[49] Harvard shamelessly harnessed the rhetorical power of statistics to convince fairgoers of the objectivity of its educational methods.

In an enthusiastic description of the exhibit, Edward Cummings, a writer for the *Harvard Graduates' Magazine,* argued that of all the displays, the section dedicated to physical training was "the crucial test of the serious purpose and scientific method of the university." The displays in this section attempted to prove that the same rigorous methods applied to the pursuit of academic knowledge could be applied to the pursuit of physical development. Included were pictures of optimal athletic facilities, books illustrating the most effective exercises, and charts recording physical measurements and "indicating defects." Any doubt as to the effective application of scientific methods, wrote Cummings, should be "dispelled by the persistent interest which men and women of all ages and conditions manifest in the typical statues of the American Student, male and female."[50] The photographs of the sculptures—representing the average college student in this context—were the focal point of the display, not samples of Sargent's charts. Photography, while not objective in the same way as numeric graphs, was considered to be the medium better suited to enhance truth values; it presented evidence in a way that was more comprehensible and compelling than even the most elaborate graph.

Displaying the sculptures as photographs in the Harvard Exhibit was a creative and resourceful way to layer multiple modes of objective representation for maximum effect. If quantification was the foundation of "true science," photography was considered the ideal vehicle of visual representation according to discussions celebrating the superiority of the photographic over the "artistic" in contemporary scientific discourse. Sargent's decision to display photographs of the sculptures, rather than the sculptures themselves, was consistent with this discourse.[51] Photographing the sculptures from the front, from the back, and in profile against a dark, neutral background imitated the photographic conventions of physical anthropology at the time, though perhaps Sargent could have been even more convincing of his objectivity had he photographed his sculptures in front of a grid or holding a yardstick, as some anthropologists did to promote comparative corporeal scrutiny.

These photographs completed a long chain of representation. The bodies were students, translated into numbers, converted into sculpture, converted into photographs. Their likenesses represented an even longer chain. They were students, captured in photographs, layered into composites, thickened into co-composites, converted into sculpture then back into photographs. They had come full circle. This elaborate sequence was an exceptional illustration of what Bruno Latour refers to as the chains of translation that make and amplify scientific truth.[52] The photographs documented bodies that did not exist alongside those that did, although visually, there was very little distinction between the two. In the context of this particular display, Sargent's sculptures were utilitarian objects that transformed the authority of statistics into the realm of visual objectivity by the ostensibly neutral eye of the camera. They made no claims to art in this display. Here the sculptures themselves were simply catalytic objects that effectively converted statistics into "disinterested" visual documentation. Like all catalysts, however, the statues were not consumed by the conversion itself.

Sargent's sculptures were given a place of honor in the Anthropology Building, where, according to the *Report of Massachusetts Board of World's Fair Managers,* they formed "the chief objects of interest and attraction" in the Laboratory of Physical Anthropology (Figure 7.6).[53] Whereas in the Harvard Exhibit, the photographs of the sculptures represented the university's faith in the "application of scientific methods," their display in the Anthropology Building involved them in a broader discourse of American nationalism. There the preferred mode of display was three-dimensional.[54] As three-dimensional objects, the sculptures would have conveyed a new kind of anthropological authority. The great proliferation of museums of natural history, anthropology, and ethnology in the late nineteenth century was predicated upon the belief that objects could tell stories even to the untrained observer. Objects, it was thought, were invaluable sources of meaning

FIGURE 7.6. Sargent's composite statues. Development Room, Hall of Anthropology, World's Columbian Exposition (1893). Gift of Frederic Ward Putnam, 1893. Courtesy of the Peabody Museum of Archaeology and Ethnology, Harvard University, PM 93-1-10/100264.1.1.

and knowledge because they could transmit information clearly and simply in a way that was not culturally or linguistically bound. The historian Steven Conn calls this assumption an "object-based epistemology," and it was the theoretical foundation of the anthropological displays at the Exposition.

Fittingly, though removed from the chaos of the Midway, the sculptures were nevertheless surrounded by an exuberant array of objects from around the world, including "relics of aborigines, cave dwellers, lake dwellers and cliff dwellers; the model homes of prehistoric man, glimpses into the domestic life of the ancient Romans unearthed from the ashes of Pompeii," including mummified bodies from Egypt and Peru, as well as a reproduction of a giant Siberian mammoth.[55] What is more, a centrally located exhibit sponsored by the government of Greece featured reproductions of the most famous ancient statuary exhumed from archeological sites, "supplemented" by contributions from the Art Institute of Chicago.[56] In short, the anthropological exhibits were a veritable smorgasbord of models, creative reproductions, works of art, relics, and contemporary objects. One might forgive any visitor who left the building without a firm grasp of what was an archeological artifact, what was a replica, and what was an imaginative interpretation, if not a complete fabrication. Like the objects that surrounded them, the sculptures of the ideal American body were meant to function as both synecdoche and metonym: they were meant to stand in for each citizen and at the same time represent the science of anthropology and the condition of national fitness. They also placed American bodies in relation to other cultures, creating an air of cosmopolitan detachment, and in relation to previous times, a sense of teleological superiority.

The statues, already raised on simple square bases, were further elevated above fairgoers on dark pedestals and individually cordoned off. Practically speaking, the pedestals and the ropes protected the delicate plaster from the crowds, but they also helped create an aura of empyrean remoteness. On pedestals, it would have been harder for viewers to relate to the sculptures physically and comparatively than if they had their feet on the ground, something that would seem to undermine Sargent's purpose for commissioning them in the first place. Displayed according to the conventions of fine art, however, they were meant to inspire. Visitors had to "look up" to the sculptures, literally and theoretically, just as they did in the Fine Arts Building where the classically inspired sculpted nude was so well represented. That they were made of inexpensive plaster would not have marked them as unfinished or artistically inferior. On the contrary, it would only have strengthened the connection between them and the display in the Court of Statuary, where many of the figural compositions were made of plaster, some originals and some casts of famous works.[57]

As Allan Wallach has noted, plaster casts were considered worthy substitutes for their originals in the last decades of the nineteenth century, and it was common practice for major American museums to become repositories of collections of plaster replicas representing otherwise unobtainable classical originals. Plaster cast collections were essential to the mission of elite college and university museums, Harvard included, where they were meant to educate and inspire their student body—and to aid in the development of a superior American society.[58] In a way, Sargent's sculptures were casts of the very students who admired the plaster replicas in their institutions' museums. When viewed in combination, the replicas in museums *and* Sargent's plaster copies, though they lacked tangible originals, went a long way in promoting aesthetic ideals, in particular classical ones.

The classicization of Sargent's sculptures was not absolute, however. Though the average woman stood in the classical poise of contrapposto and her arms were active, there were no narrative elements to explain her semi-indecent exposure. One could argue that this forthright depiction was appropriate to her statistical origins. Though she alluded to the history of the classical nude, the frankness of her pose and the placid neutrality of her facial expression concurrently would have called to mind the un-self-conscious impartiality of numbers and the seemingly disinterested presentation of scientific specimens. The male figure was similarly frank in his self-presentation. His full-frontal pose seemed to invite a kind of objective scrutiny.

Despite Sargent's ambition to spread the gospel of fitness to all Americans, regardless of sex or class, it would nevertheless be difficult not to see the exhibit of his average sculptures at Chicago's Columbian Exposition, particularly as displayed in the Anthropology Building, as a missing link in the evolutionary chain of cultural and racial being that provided the epistemological scaffolding for the fair. They linked bodies—sculpted, photographed, charted, measured, and in the flesh—that would certainly have fallen to the left or right of the average standard had Sargent submitted them to his calipers. Whether the sculptures were models of science or objects of art, a clarion call to national physical improvement or an argument for the burgeoning eugenics movement, was a matter of much debate. But it was a debate that Sargent intended to spark.

Curiously, both plaster figures were eventually painted bronze, their white bodies transformed by a subtle metallic sheen. Though it is unclear exactly when or why Sargent chose to modify his average sculptures in this way, we can speculate. The bronze color adds a kind of visual weight, a sense of solidity, that the sculptures lacked in their unpainted state. The satin glow of the bronze paint allows for the subtle play of light on the figures' contours that would have been absent on the original matte surface. Once the sculptures had served their original purpose, perhaps even Sargent himself recognized that they were no longer particularly useful as scientific models. They were, after all, conceived of as things that would actively participate in the hastening of their own irrelevance. Sargent's decision to paint the sculptures bronze arguably signaled their official retirement as objects that claimed to be simultaneously scientific and artistic and their rebirth—like so many bronze sculptures—as memorials. And yet we can still recognize traces of their brief history as pseudo-scientific objects in contemporary debates that continue to circle around images of the American body, genetics, and the myth of the anthropomorphic average and ideal.

For Sargent, the value of the sculptures lay not so much in the beauty of their forms but in the knowledge that they were three-dimensional manifestations of the vertical line that perfectly bifurcated all of his examination cards. Although the sculptures were not intended to romanticize or heroicize, Sargent's decision to commission them from Henry Hudson Kitson and Theo Alice Ruggles, artists who built their artistic careers on doing just that, demonstrates that Sargent was aware of the importance of classical aesthetics when trying to reach a broad audience and create consensus around corporeal ideals. As artistic manifestations of scientific data, Sargent hoped that the sculptures would have a unique ability to confront the public with the realities of the present and, at the same time, to contemplate future possibilities. Once on display at the Columbian Exposition in Chicago, as photographs or in plaster, the sculptures would allow statistical reality to speak for itself. They were arithmetical and aesthetic constructs that could lay claim to descriptive and prescriptive authority. Any perceived flaws in the sculptures' figures were constructive so long as they prompted visitors to the Exposition to imagine the superior American body of the future and to discuss how it might best be achieved. Even so, there is no escaping that these were difficult and troublesome objects, for while the sculptures claimed "averageness," they modeled eugenic purity and superiority even as they disavowed it. Descriptive models claim to depict concrete data—what has been or what is—while predictive models endeavor to predict future outcomes, or what will be. What is perplexing about Sargent's average sculptures is that they are both. The numbers upon which they were based were descriptive, but the way in which they were presented, as sculpture that deliberately referenced the classical tradition in style and presentation, might be better categorized as predictive. The sculptures were simultaneously testimonies and proposals.

NOTES

1 John S. Crawford, "The Classical Tradition in American Sculpture: Structure and Surface," *American Art Journal* 11 (July 1979): 38.

2 Headline from an unidentified newspaper clipping, circa 1893. Henry Hudson Kitson and Theo Alice Ruggles Kitson Papers, Archives of American Art, Smithsonian Institution.

3 Carolyn De la Peña, "Dudley Allen Sargent: Health Machines and the Energized Male Body," *Iron Game History* 8, no. 2 (2003): 3–19.

4 While Sargent worked most closely with the white middle and upper classes, confirming the racial superiority of white Anglo-Americans did not seem to be his conscious goal. On the contrary, Sargent was interested in the overall improvement of American bodies of all races, genders, and classes. To that end, he founded the Harvard Summer Institute in 1887 to train others to implement his methods far beyond the rarified college campuses of New England. To the consternation of many Harvard authorities, African Americans and women were welcome. He seemed to take pleasure in emphasizing the anti-elitist thrust of his program. The Sargent School, he wrote, had always been "possessed" by the "democratic spirit."

5 For detailed descriptions and analysis of the sculptural program at the Columbian Exposition, see Hubert Howe Bancroft, *The Book of the Fair* (Chicago: Bancroft, 1893); Norman Bolotin and Christine Laing, *The World's Columbian Exposition: The Chicago World's Fair of 1893* (Champaign: University of Illinois Press, 2002); and Pamela Potter-Hennessey, "The Sculpture at the 1893 World's Columbian Exposition: International Encounters and Jingoistic Spectacles," PhD diss., University of Maryland, 1995.

6 Frank H. Smith, *Art, History, Midway Plaisance, and World's Columbian Exposition* (Chicago: Foster Press, 1893), 2; Robert W. Rydell, *All the World's a Fair: Visions of Empire at American International Expositions, 1876–1916* (Chicago: University of Chicago Press, 1984), 56; Lee D. Baker, *From Savage to Negro: Anthropology and the Construction of Race, 1896–1954* (Berkeley: University of California Press, 1998), 57; Chaim Rosenberg, *America at the Fair: Chicago's 1893 World Columbian Exposition* (Charleston, S.C.: Arcadia, 2008); Michael Leja, "Progress and Evolution at the U. S. World's Fairs, 1893–1915," *Nineteenth-Century Art Worldwide* 2, no. 1 (2003), http://www.19thc-artworldwide.org/spring03/221-progress-and-evolution-at-the-us-worlds-fairs-18931915.

7 For good introductions to the discourse of masculinity in crisis, see Anthony Rotundo, *American Manhood: Transformations in Masculinity from the Revolution to the Modern Era* (New York: Basic Books, 1993); Michael Kimmel, *Manhood in America: A Cultural History* (New York: Free Press, 1996); Gail Bederman, *Manliness and Civilization: A Cultural History of Gender and Race in the United*

States, 1880–1917 (Chicago: University of Chicago Press, 1996); Athena Devlin, *Between Profits and Primitivism: Shaping White Middle-Class Masculinity 1880–1917* (New York: Psychology Press, 2005); John Kasson, *Houdini, Tarzan, and the Perfect Man: The White Male Body and the Challenge of Modernity in America* (New York: Hill and Wang, 2002); and Dana D. Nelson, *Capitalist Citizenship and the Imagined Fraternity of White Men* (Durham, N.C.: Duke University Press, 1998). For a more general overview, see Herbert Sussman, *Masculine Identities: The History and Meanings of Manliness* (Santa Barbara, Calif.: Praeger, 2012), and Ruth Marie Griffith, *Born Again Bodies: Flesh and Spirit in American Christianity* (Berkeley: University of California Press, 2004); see also Martha H. Verbrugge, *Able-Bodied Womanhood: Personal Health and Social Change in Nineteenth-Century Boston* (New York: Oxford University Press, 1988), and James Whorton, *Crusaders for Fitness: The History of American Health Reform* (Princeton, N.J.: Princeton University Press, 1982).

8 See George Miller Beard, *American Nervousness, Its Causes and Consequences* (New York: Putnam, 1881). Horatio C. Wood, a prominent physician, complained that "the exigencies of modern life are producing and ever-increasing amount of nervous disease." See Horatio C. Wood, *Brain Work and Overwork* (Philadelphia: P. Blakiston, 1885), 94. Historians have drawn attention to the links between neurasthenia and its troubling relationship to contemporary discourses of class, race, gender, and modernization. See John S. Haller and Robin M. Haller, *The Physician and Sexuality in Victorian America* (Carbondale: Southern Illinois University Press, 1995), 5–43. The historian Carolyn Thomas de la Peña has linked (predominantly white) Americans' ideas about technology and energy, so essential to modernization, to the national conversation about fitness and the ideal body. Julian B. Carter has argued that the widespread diagnosis of neurasthenia was racially coded. See De la Peña, *The Body Electric,* xiv, and Julian B. Carter, *The Heart of Whiteness: Normal Sexuality and Race in America, 1880–1940* (Durham, N.C.: Duke University Press, 2007), 43.

9 Dudley Allen Sargent, "The System of Physical Training at the Hemenway Gymnasium," in *Physical Training: A Full Report of the Papers and Discussions of the Conference Held in Boston in November, 1889,* ed. Isabel C. Barrows (Boston: Press of George H. Ellis, 1890), 65.

10 Sargent outlined the details of his examination methods in his pamphlet *Measuring and Testing the Principle Physical Characteristics of the Human Body* (Cambridge, Mass.: Harvard University, 1887). Other guides to the science of anthropometry were available at the time, including Jay W. Seaver's *Anthropometry and Physical Examination: A Book for Practical Use in Connection with Gymnastic Work and Physical Education* (New Haven, Conn.: privately printed, 1896). Most of these cards are contained in the Dudley Allen Sargent Papers in the Harvard University Archives.

11 Dudley Allen Sargent, *An Autobiography* (Philadelphia: Lea and Febiger, 1927), 151.

12 Thanks to Sarah Wasserman for this important point.

13 Dudley Allen Sargent, "The Physical Proportions of the Typical Man," *Scribner's*, July 1887, 3–17, 6.

14 Sargent does not mention the exact date when he compiled his data and came up with his average standards, though he says in 1887 that it was the result of seventeen years of work and incorporated measurements from men ages seventeen to thirty. The female standards were developed with the help of "trained assistants at the principle female colleges." He describes the process in "Physical Proportions of the Typical Man," 11.

15 Kasson, *Houdini, Tarzan, and the Perfect Man*, 41.

16 Sargent, "Physical Proportions of the Typical Man," 16.

17 There are subtle and important distinctions to be made, correspondences to acknowledge, and plenty of semantic slippages between an "average" or "mean," a "type" or "the typical," and the "normal." Sargent's own language attests to this linguistic tangle. Elizabeth Stephens and Peter Cryle have convincingly argued that the notion of "normalcy" was an expansive concept that gained greater and greater traction in the United States as the twentieth century progressed. The term *normal* was "largely confined to professional discourses" until the last decade of the nineteenth century but was virtually synonymous with "average" and "typical" by the middle of the twentieth. Until then, normality "remained an emergent, rather than an established idea in popular culture." See Stephens and Cryle, "Eugenics and the Normal Body: The Role of Visual Images and Intelligence Testing in Framing the Treatment of People with Disabilities in the Early Twentieth Century," *Continuum* 31, no. 3 (2017): 365–76, and their chapter "The Object of Normality" in *Normality: A Critical Geneology*, 294–332 (Chicago: University of Chicago Press, 2017).

18 De la Peña, *The Body Electric*, 64; Carolyn De la Peña, "Dudley Allen Sargent: Health Machines and the Energized Male Body," *Iron Game History* 8, no. 2 (2003): 3–19.

19 *New York World*, "Strongest Man in the World," June 18, 1893, 16.

20 *New York World*, June 25, 1893, 21.

21 Ellery Foutch, "Arresting Beauty: The Perfectionist Impulse of Peale's Butterfly's, Heade's Hummingbirds, Blaschka's Flowers, and Sandow's Body," PhD diss., University of Pennsylvania, 2011, esp. chapter 3, "Embodying Perfection: The Petrification of Eugen Sandow," 167–284; Michael Squire, *The Art of the Body: Antiquity and Its Legacy* (New York: Oxford University Press, 2011), 16–18.

22 See Kasson, *Houdini, Tarzan, and the Perfect Man*, 54. An advertisement for Sandow's Boston school noted that the Sandow cast "had been duplicated in bronze for Harvard University, where it may now be seen." "Sandow Will Mail

Free," *Public Opinion* 32, no. 16 (1902): 511; Foutch, "Arresting Beauty," 265n278. No photographs of this bronzed cast survive, and its whereabouts are unknown. I suspect that Sargent had his casts of the average man and woman painted bronze around 1902 so that the three sculptures would make a complementary trio.

23 Dudley Allen Sargent, "Physical Proportions of the Typical Athlete," *Scribner's*, November 1887, 3–17.

24 Unidentified clipping, Henry Hudson Kitson and Theo Alice Ruggles Kitson Papers, Archives of American Art, Reel 3930.

25 Dudley Allen Sargent, "The Harvard Summer School of Physical Training: Its Aims, Its Methods, and Its Work," *Boston Medical and Surgical Journal* 134 (February 20, 1896): 184.

26 For an in-depth analysis of the how the sculpted bodies at the Columbian Exposition contributed to and complicated the spectacle of the fair, see Diane Dillon's "The Fair as Spectacle: American Art and Culture at the 1893 World's Fair," PhD diss., Yale University, 1994.

27 An article in the *Illustrated American* attributed the average female sculpture to Ruggles and the average male sculpture to Kitson. I have not found other archival evidence to suggest that the division of labor was so clear. Milton J. Stone Jr., "The Typical Man and Woman of America," *Illustrated American*, June 17, 1893, 709–10.

28 Kitson studied sculpture under Jean-Marie Bonnassieux at the École des Beaux-Arts in Paris. In 1888 he returned to the United States to establish his own successful studio. He was the only American sculptor at the Exposition Universelle in Paris to receive a medal for his work.

29 Ruggles specialized in war memorials at a time when soldier monuments were becoming increasingly popular. See "Sentries, Doughboys, and GI Joes," in Donald Martin Reynolds, *Masters of American Sculpture: The Figurative Tradition from the American Renaissance to the Millennium*, 153–67 (New York: Abbeville Press, 1993), and Jennifer Wingate, *Sculpting Doughboys: Memory, Gender, and Taste in World War I Memorials* (Burlington, Vt.: Ashgate, 2013).

30 Squire, *Art of the Body*, 8–15.

31 A general lack of interest in statistical measurements did not mean that there was not an interest in scientific, medical, or photographic constructions of the American body. Thomas Eakins was foremost among artists who promoted such representation in painting. His famous *The Agnew Clinic* (1889) was shown at the Columbian Exposition and sparked much criticism and controversy for its scientific realism. Eadweard Muybridge's photographic studies of animal and human locomotion and zoopraxiscopes were also on display at the Exposition.

32 See also Stone, "Typical Man and Woman of America," 710.

33 David Bindman, *Ape to Apollo: Aesthetics and the Idea of Race in the 18th Century* (Ithaca, N.Y.: Cornell University Press, 2002), 92–100.

34 Unidentified clipping, Henry Hudson Kitson and Theo Alice Ruggles Kitson Papers, Archives of American Art, Reel 3930.

35 In a particularly ambitious photographic project, Allen and Rowell combined 449 portraits of American male college students, most from Harvard, and created a complementary composite of the American college female from 287 portraits. Phillip Prodger, *Darwin's Camera: Art and Photography and the Theory of Evolution* (New York: Oxford University Press, 2009), 217. These photographs were published in John T. Stoddard, "College Composites," *Century Illustrated* 35, no. 13 (1887): 121–26.

36 Karl Pearson, *The Life, Letters and Labours of Francis Galton* (London: Cambridge University Press, 1914), 2:284.

37 John T. Stoddard, "Composite Photography," *Century Illustrated* 33, no. 5 (1887): 750–58.

38 Francis Galton, "Composite Portraits, Made by Combining Those of Many Different Persons into a Single Resultant Figure," *Journal of the Anthropological Institute of Great Britain and Ireland* 8 (1879): 134.

39 Galton, 132.

40 W. Lincoln Adams, "Composite Photography," *Outing, an Illustrated Monthly Magazine of Recreation,* April 1891, 31–33.

41 Allan Sekula, "The Body and the Archive," *Art Journal* 41 (Spring 1981): 15–25.

42 The composite photographs of college classes were later featured in an official publication of the Second International Exhibition of Eugenics held in New York in 1921. Kris Belden-Adams, "'Composita,' the 'Mascot' of the Smith College Class of 1886: Picturing Sisterhood, Social Castes and Gender Roles," in *Photography and Failure: One Medium's Entanglement with Flops, Underdogs and Disappointments* (New York: Bloomsbury Academic, 2017), 109.

43 Galton, "Composite Portraits," 134.

44 Sargent wrote in an undated letter to Harvard president Charles Eliot that he contributed to the World's Fair display "two life size statues of the typical male and female college students, made from 10,000 measurements." HUG 1768.2 Sargent, D. A. Correspondence, Financial Records and Personal History.

45 *Illustrated American*, "History of Seven Days," April 15, 1893, 465.

46 Amy Robsart, "Columbia and Jonathan," *Boston Post,* March 21, 1893.

47 Stone, "Typical Man and Woman of America," 708; *Boston Sunday Globe,* "Tale of Art and Love," June 25, 1893.

48 For more on the unveiling strategies of public monuments in the nineteenth century, see Kirk Savage, *Standing Soldiers, Kneeling Slaves: Race, War and Monument in Nineteenth-Century America* (Princeton, N.J.: Princeton University Press, 2018); Joy M. Giguare, *Characteristically American: Memorial Architecture, National Identity and the Egyptian Revival* (Knoxville: University of Tennessee Press, 2014). In *Memorial Mania: Public Feeling in America* (Chicago: University

of Chicago Press, 2012), Erika Doss pushes the discussion of monuments and memorials into the twentieth and twenty-first centuries in ways that both link and break from the nineteenth-century tradition.

49 Edward Cummings, "The Harvard Exhibit at the World's Fair," *Harvard Graduates' Magazine,* September 1893, 50–63, 50–51, 55.

50 Cummings, 61.

51 For an enlightening examination of the role of photography in the context of the Columbian Exposition as a whole, see Julia K. Brown, *Contesting Images: Photography and the World's Columbia Exposition* (Tucson: University of Arizona Press, 1994), particularly chapter 3, "Appropriating the Image: Using Photographs for Display." See also Massachusetts Board of Managers, *Report of Massachusetts Board of Managers World's Columbian Exposition, 1893* (Boston: Wright and Potter, 1894), 113.

52 Bruno Latour, "Drawing Things Together," in *Representation in Scientific Practice,* ed. Michael Lynch and S. Woolgar, 18–68 (Cambridge, Mass.: MIT Press, 1990).

53 Massachusetts Board of Managers, *Report of Massachusetts Board of World's Fair Managers,* 160.

54 Massachusetts Board of Managers, 31; Steven Conn, *Museums and American Intellectual Life, 1876–1926* (Chicago: University of Chicago Press, 1998), 75–113.

55 Bolotin and Laing, *Chicago World's Fair of 1893,* 77.

56 The Greek exhibit included, for example, statues of Diana, Apollo, and a number of allegorical figures. Bancroft, *Book of the Fair,* 634.

57 For a list of the sculptures on display in the Fine Arts Building at the Exposition, see Moses Purnell Handy, ed., *World's Columbian Exposition, 1893, Official Catalogue* (Chicago: W. B. Conkey, 1893), 10–13.

58 Alan Wallach, "The American Cast Museum: An Episode in the History of the Institutional Definition of Art," in *Exhibiting Contradiction: Essays on the Art Museum in the United States,* 38–56 (Amherst: University of Massachusetts Press, 1998).

8

Hypermodels: Architectural Production in Virtual Spaces

SEHER ERDOĞAN FORD

AN ARCHITECTURAL HERITAGE SITE OF DUAL COMPLEXITY

Digital reconstructions are the last and best chance to document and, in a sense, preserve imperiled sites of architectural heritage. But there is a conflict at their core. On one hand, they are often the result of layers of interpretation put forward by different scholars of fragmentary remains and incomplete, sometimes contradictory records. On the other hand, they are built on software platforms that demand discrete and unambiguous data, such as precise dimensions and specific material designations. Moreover, such platforms are designed to represent the final state of an artifact; the decisions and the degrees of certainty that underpin the data, designs, and decisions surrounding them are lost, as is the possibility of describing significant alterations that may have occurred over the life of a building. In what follows, I discuss the "hypermodel" as a form that aims to recover the process by which digital reconstructions are made and, in doing so, also reveal its contours as a model. The hypermodel makes visible, and manipulatable, not just the representation of a building, including its history of modifications, but also the entire chain of scholarly reasoning and judgment that goes into making a model. By considering a hypermodel that I have helped construct, this essay explores how digital, 3D representations can function as dynamic tools that account for interpretation while reconstructing architectural sites. Furthermore, the discussion suggests how such reconstructions are capable of staging a more-than-textual act of "reading" architectural sites through the multifaceted lens of material culture, conveying the interrelations between materials and objects, buildings and people.

The need for such a model emerged during my study of the site of Studious Church, later known as the Mosque of Imrahor, in present-day Istanbul, Turkey. The oldest surviving ecclesiastical Byzantine structure in Istanbul, the church of St. John Studious Monastery was erected around A.D. 450 by the Roman senator

Studious, possibly on the grounds of an existing parochial church, near the "Golden Gate" situated along the Theodosian city walls delineating Byzantium of the time.[1] Initially, the monastery housed the order of Acoemetae, otherwise known as the "sleepless monks," who characteristically held their services continuously, day and night. The monastery began to gain larger prominence in the Roman Catholic community in the eighth century, when as many as seven hundred monks resided within its bounds.[2] The Byzantinist scholar Alexander Van Millingen describes the monastery in the eleventh century as a refuge for devoted monks, overexposed emperors, and even dissidents—and as a focal point of clerical and intellectual culture in the Roman world.[3] The *Catholic Encyclopedia* also tells us that the site was a burial ground of holy bodies, the center of Byzantine religious poetry, and an educational community for calligraphy.

The monastery was in a state of complete neglect following the siege of Constantinople in 1204, when the site ostensibly became a grazing ground for sheep. Upon the Byzantine Empire's demise in 1453 and the Ottoman settlement of Constantinople, all liturgical functions of the monastery ceased. The larger complex was abandoned and left to disintegrate, with only the basilica-plan church structure and some of its ancillary areas remaining intact. Until the early sixteenth century, the grounds were used as a stone yard, serving the Ottoman sultan's construction projects throughout the city.[4] Around 1485, the sultan's equerry, a high-ranking officer in charge of the stables, converted the monastery church to a mosque, thereby inspiring the name "Imrahor," which means "the Stablemaster." Subsequently, the mosque was designated a Sufi dervish lodge, or *tekke,* dedicated to the brotherhood of the Sunbuli, a branch of one of the most prominent Islamic sects in the Ottoman Empire.

The most significant present-day milestone for the site was a renovation project undertaken in 2016 by the Directorate General of Foundations administration, which has had jurisdiction over the property since the early 1920s, coinciding with the transition from the Ottoman Empire to the Turkish Republic and the abolishment of all *tekkes.* Even though the building was made a museum by the administration in 1946, the site has not been open to the public since and has fallen into decay. Though roofless and exposed to the elements, invaded by vegetation, it continues to attract tourists undertaking Eastern Orthodox pilgrimages as well as scholars. Current construction plans include rebuilding the mosque, which promises a politically charged outcome.[5] Word of these plans has already raised controversy among various constituents, including the local and displaced Greek population, who continue to regard the basilica as sacred, and others who criticized the lack of information shared about the renovation plans.[6]

In addition to its historical significance as a religious, cultural, and education center in the Byzantine Empire and a prominent center for the Sunbuli sect of

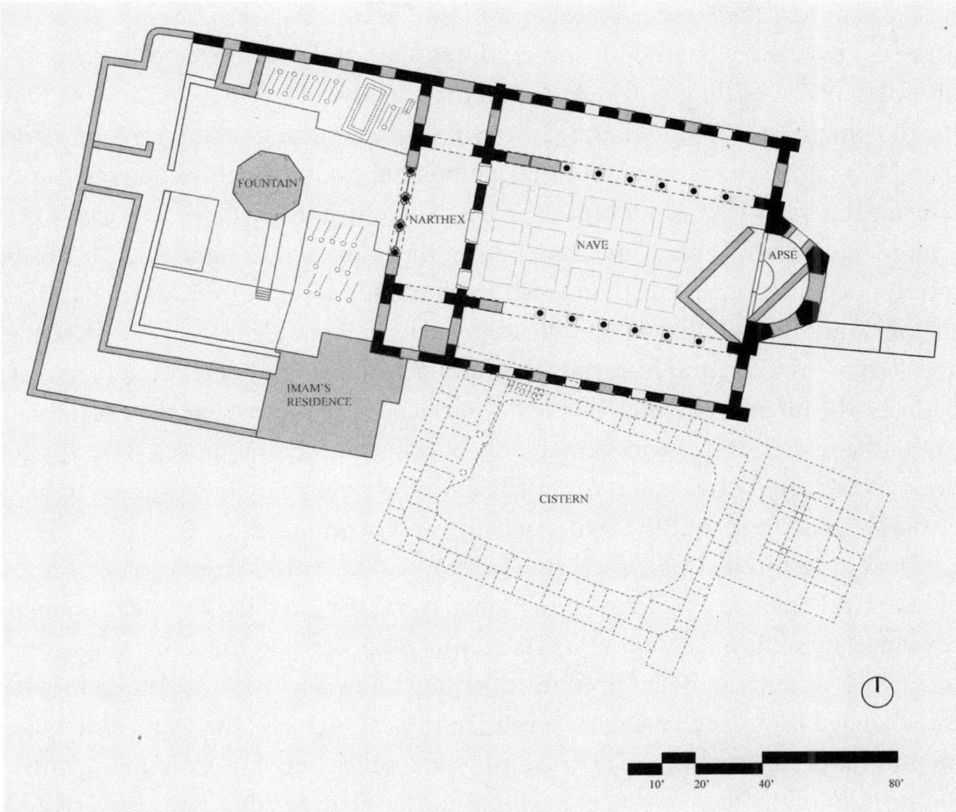

FIGURE 8.1. A diagrammatic layout of the Church of Studious as converted to the Mosque of Imrahor. The black fill indicates the general extent of the church structure, while the gray fill locates the modifications undertaken during and after the conversion. Image by the author.

Sufis during the Ottoman reign, Studious is a site of "dual complexity" in terms of architectural production. First, it was built up and altered over sixteen centuries. Multiple rounds of major reconstruction were conducted in response to natural disasters, such as fires and earthquakes, and changing functional and aesthetic agendas, on top of routine, small-scale maintenance recorded in Ottoman state archives.[7] Second, the ways in which the building has been visualized have evolved over time, resulting in multiple and often incommensurate representations of the same site (Figure 8.1).

An early example of the early Christian basilica, the original church was nearly square, approximately 26 by 34 meters (85 by 112 feet). Most of the building-scale architectural elements are still discernable from the remaining structure, including the three-bay nave following the conventional east–west orientation, with galleries over both side aisles; a semicircular apse at the interior, expressed in three facets on the exterior; a pillared narthex; and an underground cistern.[8] Many elements, however, have not survived, including the timber roof over the central nave. We know from Van Millingen's research that there are two likely possibilities for how the roof might have appeared. Based on writings and drawings, this nave roof would have been higher than that of the aisles, comparable to the roughly contemporaneous basilica-plan Eski Juma Mosque. Alternatively, it might have had clerestory windows like Saint Demetrius Church.[9] Another vanished element is the intricate interior adornment that distinguished the Studious Church from other basilicas that were hastily built throughout the Roman world in the Near East, North Africa, and Italy.[10] Architectural historians, including scholars of architectural ceramics, describe the interior of the monastery as surfaced with marble panels, Corinthian capitals, and architraves with remarkable polychrome vegetal motifs. The marble floor, constructed of *opus sectile,* an ancient inlay patterning technique featuring elaborate geometric motifs, is still partially present on the site.[11]

In addition to its original merits, Studious today is the product of centuries of historical change, a palimpsest of many layers that are difficult to disentangle. Evidence for some of these overlapping layers is apparent in the remaining structure, while others can be inferred from written and visual records. When the site was changed for use as a mosque, several modifications were made, including the incorporation of more of the area around the basilica and a near-doubling of the footprint. A minaret; *şadırvan,* or pavilion-like fountain for ablution practices; and a residence of an imam, the worship leader of the mosque, were added. A smaller, shallower apse dome with new apertures was reconstructed. Additional changes related to liturgical practices impacted the site at a smaller scale, including new partitions in some of the colonnade openings, infilled windows, and doors. The most symbolically and physically prominent aspect of the conversion to a mosque involved the reconfiguration of the *templon,* the boundary that separates the nave from the altar, and the insertion of a *mihrab,* an architectural niche on the apse wall, to reorient the building toward Mecca. This reorientation, thirty-three degrees off the east–west axis, was accomplished by moving fifth-century marble strips, along with brick and stone fragments, to suggest the new direction (Figure 8.2).[12]

Studious is extraordinary in terms of the temporal scale it encompasses and for the way it embodies a layered history. But what makes it even more compelling is that it exposes more acutely the processes that countless buildings commonly undergo. Cultural anthropologist Tim Ingold elaborates on the production of architecture in his book *Making,* arguing that such production extends in time and space far beyond initial assembly of building components. It also includes the extraction of natural materials; the fabrication of components and their transportation to the site for assembly; changes over time, including additions and removals; and even

FIGURE 8.2. Photograph taken in 1937 by historian and explorer Nicholas V. Artamonoff capturing the amalgamation of building materials at the scale of a wall detail. Courtesy of the Nicholas V. Artamonoff Photographs of Istanbul and Turkey, 1935–1945, Dumbarton Oaks, Trustees for Harvard University, Washington, D.C.

its recycling into other forms and uses. Viewing the life of the building along a continuum—what I refer to from here onward as its "building-life"—entails a shift in scholarly perspective, as it suggests that the act of building is never finished. The concept of building-life expands the narrow focus from the *product* to an enlarged scope encompassing the *process* of growth, decay, and regeneration. Such conceptual reframing and its representation that would interrelate physical attributes of building materials with their active roles in social dynamics can uncover historical practices that ultimately reveal the contours of change not only at the scale of the artifact but also at a societal level. Therefore, following Ingold's lead,

I take the building-life of Studious as a case study and motivation for a new form of architectural representation.

Just as Studious was physically remade over more than one and a half millennia, its visual representation also evolved dramatically. This evolution entailed changes in technique and media, differing purposes behind each image, varied authors' agendas, and different degrees of access to the site over time. Given the long history of the building, there are also chronological gaps in-between when archival images were made. In viewing the entire collection of images made of the Studious Church, one quickly sees that many of the visual representations of the site take an interior view of the apse (and later where the *mihrab* is placed) as their subject. The photographic records from different sources situate the viewer in the same space, but at different points in time. In a sense, they render cycles of growth, decay, and alteration referred to by Ingold. One photograph, for example, perhaps one of the earliest and most cryptic of the group, was taken when the building was in ruins—missing its roof and walls falling apart—but still being used for prayer by the Sunbuli community members (Figure 8.3). The exact provenance of the image is uncertain.[13] But based on what we know from other dated records, we can deduce that this photo must have been taken before the 1920s, right around the time when the building was deemed a nonreligious government asset. The fragmented surfaces depicted in the photo, including the plaster treatment applied by the Ottomans on Christian iconography peeling off, and receding background figures palpably evoke a sense of disintegration. Another photograph belongs to a collection of the photographer, historian, and explorer Nicholas V. Artamonoff, who set out to document the remnants of Byzantine architecture across what was previously Ottoman land.[14] In these images, architectural finishes are completely eradicated, and the area surrounding the apse appears barren, except for the intermittent appearance of the imam. A third set of photographs, more analytical in character, was made by researchers with the German Archaeological Institute during their field study in the 1930s.[15] In these instances, the building appears devoid of human activity (except for occasional visitors), with burgeoning vegetation covering the apse and the *mihrab* to the point of obfuscating both. A further set of photographs shows continued disuse, with vegetation everywhere, but with some scaffolding set up in anticipation of restoration efforts. This is the version captured by a team from the New York University Institute of Fine Arts, led by Professor Thomas Mathews in the early 2000s.[16] Taken together, these photographs span the twentieth century and compose a "time lapse," a sequence of points along a wide temporal arc into the same architectural space showcasing at once the site's different material states and the blurriness of the divide separating its cultural and material delineations within.[17]

FIGURE 8.3. Photograph of the apse depicting the main space around the 1920s, when the building was still partially occupied as a mosque despite its ruinous state. Courtesy of Fatih Köse as published in his article "Imrahor Ilyas Bey Camii ve Osmanli Doneminde Gecirdigi Tamirler," *Vakif Restorasyon Yilligi* 4 (2012): 37.

Another type of difference among representations of Studious is *spatial*, as seen in the juxtaposition of two images in particular. Completed independently within a few years of each other, one print shows the building as a church and the other as a mosque. The church version is an etching by architect Wilhelm Salzenberg published in 1854.[18] The depiction as a mosque is by the Greek poet Antonios Manarakis.[19] Commissioned by King Frederick Wilhelm IV, Salzenberg visited Istanbul numerous times from 1847 to 1867 and published *Old Christian Architectural Monuments of Constantinople,* which features his original drawings. In this set of technical drawings, including plans, sections, and elevations, Salzenberg portrays Christian monuments of Istanbul. In the case of the Church of Studious, his drawings involve interpretive reconstruction. The drawings give some indication of materiality through hatching and shading on the exterior of the building, while the interior is left mostly blank. There are no people in Salzenberg's drawings. In contrast, Manarakis made several watercolors that represent the building in its state as a mosque. One view shows the interior space used by presumably the Sunbuli community. Because we do not know who commissioned the drawings or why Manarakis made them, it is hard to know how he gained access to the building. However, given the amount of detail, it appears he drew from observation. His watercolors include vivid details of the interior, suggesting how the Sunbulis modified and occupied the space, signaling the notion of the physical artifact as not only an active participant but a formative agent in the cultural practices that it hosted.[20] Taken as a pair, Salzenberg's and Manarakis's images demonstrate how representations made at the same moment can illustrate wholly different ideas of the same building. A hypermodel would need to be able to represent both states, however divergent they may be, as well as the conditions that motivated each of the different representations. These examples reinforce the idea that a site like Studious, with an extended history and multiple physical, representational, and conceptual versions, requires a mode of representation that can situate its variability over time and in regard to diverse constituencies.

THE INADEQUACY OF CURRENT MODELS OF REPRESENTATION

Current forms of modeling historically complicated sites all too often fall short. For example, Van Millingen created a set of diagrammatic drawings that indicate repairs dating to the pre- or postrepublic eras, the latter labeled "Turkish," with the understanding that other areas were to be read as "Byzantine." While useful in conveying the building's extended life beyond the Byzantine liturgy, Van Millingen's nomenclature is clearly driven by contemporaneous politics; it draws abstract boundaries and creates synthetic breaks in what are actually intertwined components and amalgamated eras of an evolving building. Most Ottoman interventions used on-site materials and involved surgical tactics to infill, cover, replace, and so on.[21] The notion of a "Turkish hand," as Van Millingen implied, imposes a reductive narrative, flattens layers, and defaults to a "before-and-after" view of authorship at the expense of a continuous and complex reading of the building-life of the artifact.

In recent years, there have also been a number of historical reconstruction projects using a variety of digital modeling platforms and techniques.[22] At first glance, digital media might seem better suited to represent the overlapping layers and dynamic boundaries of a site like Studious. But conventional uses of digital media often trend in the opposite direction, often aiming to converge on a singular likeness at high resolution. Where nebulous effect is applied in digital representations of buildings, it is often done toward stylistic or aesthetic aims rather than toward engendering a multiplicity of meaning. Digital 3D models of architectural heritage sites are typically made to represent what we think buildings looked like at a specific moment. Materials, lighting conditions, and geometries are calibrated for precision and vivid portrayal.

The emphasis on realistic and determinate building models crowds out the possibilities for ambiguity and interpretive flexibility, eliding the complexity of a site like Studious. For example, the objective of the digital reconstruction project *Byzantium 1200* is to recapture significant monuments and the overall architectural character of Byzantium in the year 1200.[23] To this end, buildings are rendered photorealistically, even though the project authors point out that signs of aging and urban decay, which would have been clearly evident at the time, have been erased in their model. The digital model of *Byzantium 1200* presents a Byzantium that never was; it offers a "realistic" depiction of a fantasy. Rather than providing such airbrushed reproductions, no matter how realistic, digital models might be more robustly understood as resources for interpretation, capable of capturing the effects of time, intention, perspective, and partial knowledge.

A NEW FORM OF REPRESENTATION: THE HYPERMODEL

In his writings, *Translations from Drawing to Building,* architect and historian Robin Evans discusses the relationship between the original idea, its representation, and the realized building.[24] He characterizes the movement from one point to another as an act of translation through media that yield conceptual friction and, in turn, altered intentions. Evans argues that such transformation is the virtue—if not the purpose—of representation. A drawing's or model's *unlikeness* to its referent signifies discovery and therefore meaning. Each stage in the transformation leaves a gap for divergences from an original concept, opening vistas on to new possibilities and, thus, the potential for new meanings.

In Studious, the transformational steps are many, occurring at many scales and over centuries. Taking into account the different forms of representation, as well as the varied perspectives of the building's users, the steps become innumerable. Likewise, the gaps between the planned changes and the physical results open up to incalculable possibilities. A hypermodel of Studious would not only trace each of these stages of building change; it would also make available to researchers the moments when design ideas change into representations that are then translated into actual interventions in the building's fabric. Therefore the digital representation of a building requires a dynamic medium, in which multiple forms and heterogeneous media can relationally exist. What this describes is an interactive and spatial environment, preferably occupiable or inhabitable in some fashion. Virtual reality (VR) is capable of creating such environments.

The National Center for Supercomputing Applications defines VR as "a medium composed of highly interactive computer simulations that sense the user's position and replace or augment the feedback of one or more sense—giving the feeling of being immersed."[25] This suggests that for historical reconstruction research, the transition from a digital three-dimensional modeling environment to the immersive VR environment entails not only a spatial but a sensory leap. VR allows for navigation within the digital space of the model, affording agency delivered through free movement as well as haptic experience that engages the full range of perception, such as balance and temperature, and a sense of touch localized in the skin. The VR user becomes a visitor, a navigator, and an agent of interaction and is in turn reminded of her sensorial relationship to the architectural space she virtually accesses. As a form of modeling, VR could veer, in the spirit of existing high-resolution 3D representations, toward fantastic, and fantastical, replications. Alternatively, a VR model that emphasizes processes and stages of transformation could become a powerfully critical tool for historical understanding.

Perhaps counterintuitively, the relationship between the digital model and the virtual visitor's body does not necessarily restrict VR to full-scale representation. On the contrary, representation across multiple scales can be central to virtual, immersive reconstructions. As architect Paul Emmons argues, 1:1 scale is not the ultimate metric behind an architectural drawing. Scaled drawings facilitate comprehension of the designer's intentions, the imagined form or model.[26] For this reason, scaled representation can be a virtue for VR. Drawings within the virtual environment can lead the virtual visitors' imagination beyond what they are seeing and moving through.

In the 1990s, media scholar Pierre Lévy coined the term *hypermedia* to define a virtual reservoir that contains images, text, sounds, and even tactile qualities and is connected to external content.[27] A number of contemporary applications within the broader context of digital space allude to the term *hyper-*. Perhaps the best-known example is hypertext, in which a piece of text becomes a portal to related information, both textual and visual, that is accessed by clicking on it with a mouse or similar instrument. Other uses of the term *hyper-* attest to its hybrid and boundary-breaking character agenda.[28]

Recently, several colleagues and I tried to build a hypermodel, following the ideas described here and based on our research of Studious.[29] It presumes a spatial construct that hosts heterogeneous content in terms of media type and scale and in degrees of reliability, that is, in terms of the sources of information. Its virtual environment facilitates not only immersive spatial navigation but also temporal exploration. We conceived of it as a self-transforming platform capable of conveying changing aspects of Studious over time and deepened by the extensions—links, so to speak—to other content embedded within various parts of the model.

Our team included a VR technologist, an interaction designer, an architectural designer, and several students of architecture. Together, we digitally modeled the Studious building across various platforms, including Rhino, 3D Studio Max, and SketchUp; developed the VR model using the software application UnReal Engine; and tested the hypermodel experience using the HTC Vive VR headset. Our objective throughout our collaboration was to develop a hypermodel prototype, one that offered a new method of representation and addressed the complexity of Studious and its representational history in appropriately nuanced, dynamic, and evocative ways.

When a visitor "enters" the hypermodel, she encounters a digital construct with its own graphic and experiential language, rather than a naturalistic simulation of the Studious building. The message that "this model is not equal to that building" is foregrounded. The design team emphasized visual and conceptual distance, or purposeful "unlikeness," by employing a limited, abstracted material palette, with desaturated finishes and textures. For instance, interior columns are rendered in a muted, low-resolution stone texture, similar in color to but less saturated than their historical counterparts, which were a rich green, verde antique stone.

As the visitor explores the hypermodel in VR space, she sees a sunny sky through the windows and moves in and out of shadows, such that the real-world experience of light perception, with its concurrent sensations of contrast, glare, and dimly lit spaces, is simulated. She is free to look and move in all directions in the virtual space. However, as is common with VR, devices in the physical space delineate the boundaries of the virtual space. The boundary appears as a blue mesh if the visitor walks up to it, and the virtual model abruptly disappears if she tries to cross it. In addition to free movement within the boundary, VR affords a functionality called "teleporting," by which the visitor can leap from one location to another in the virtual model. The current technical capacity of VR platforms affords continuous movement in smaller areas and discontinuous navigation throughout a model.

While moving through the Studious hypermodel, one encounters certain elements of the building, which are highlighted in bold, pulsing blue outlines that stand apart from the muted background and signal interactivity (Figure 8.4). Embedded within these building elements is an array of information, accessed via textual and multimedia "annotations." The annotations are floating, opaque surfaces at eye level, arranged in a shallow arc for convenient reading. Examples of these highlighted elements include a typical column along the basilica colonnade, a typical window in the aisles, and the dome in the apse. Careful considerations of the appropriate density of information and a desire to avoid oversaturating the virtual model informed our decisions regarding the number and specific locations of annotated elements. There are two primary ways supplementary information can be displayed in any VR environment: as panels that appear independently of the virtual model, fixed within a view frame, or at specific coordinates within the virtual model and disassociated from the visitor's view. We opted for the latter "in-world" annotations, associating them with architectural elements to reinforce the relationship between the annotative contents and their referents. This relationship makes the experience more intuitive, since annotations respond to proximity: visitors can choose to engage with them or, by moving away, cause them to disappear.

The embedded annotations in the Studious hypermodel contain selected archival material and original visualizations created by the design team. The archival items are especially valuable now that the site has been closed to visitors since 2016; field investigation, including photography, is now forbidden by the Turkish government.[30] In addition, an audio narrative is included to further explain key elements. Like the other annotations, audio begins and ends in relation to the proximity of the visitor. All of the annotations highlight the building's transformation over time.

Within each annotation panel, a thumbnail image appears to the left; text, with hyperlinks, is in the middle; and the right section offers browsable links to further content, including additional representations at different scales, ranging from material close-ups to technical drawings to overall views. The range of available scales and image functions means visitors can analyze discrete elements in minute detail or appreciate broad perspectives. Whether still or in motion, the visitor can access multiscalar information through annotations and thereby gain the opportunity to connect the finest, smallest material scale to the largest architectural or conceptual scale. In addition, nested within each representational tier are pairs or trios of images portraying the column at different points in time. In one case, for example, the visitor sees a black-and-white photograph published by Van Millingen around 1910 next to a color photograph of the same column capital from the 2000s. Made almost a century apart, these two images bookend the transformation of an intricately painted column capital to an eroded stone mass. In this way, the hypermodel presents a constellation of references that create a sense of a dynamic building-life and that foregrounds not just the variability between representations but their fundamental and at times unbridgeable differences. Additionally, the annotations create multiple entry points for object analysis at biographical and genealogical scales, which can be interrelated with the social practices, rituals, and craft traditions in changing cultural contexts.[31]

FIGURE 8.4. *Top to bottom,* the still images capture the visitor's experience interacting with the annotated elements embedded within the hypermodel. Image by the author.

In addition to the annotations attesting to the rich building-life of Studious, the hypermodel offers visitors experiential modes. "Temporal layering" is one such mode that rehearses an animated history of changes as the visitor inhabits the virtual space. In one sequence, for example, a visitor would witness how the volume, light, and details of the apse and dome overhead changed from the eighth-century Byzantine design to its sixteenth-century Ottoman state. Some states were based on documents or material fragments, whereas others, because of a lack of evidence, were interpolated. Each stage is treated as a separate layer with a distinct graphic style that indicates the degree of scholarly certainty regarding the reconstruction. The eighth-century dome is rendered diagrammatically to indicate our approximate understanding of the view at that time (Figure 8.5). A subsequent wireframe depiction foregrounds what we know of the dome's geometry as it shifted toward its Ottoman-era expression, smaller overall and lower, with some windows lost or lowered and with the reoriented *templon* situated in front of the apse. The animated sequence continues as the wireframe layers recede and the solid masses, now rendered, of the Ottoman structure become clearer and more prominent. With this sequence, the hypermodel connects an immersive, almost visceral experience of formal and material transformation with whatever analytical grasp visitors may have gained through annotations.

STUDIOUS HYPERMODEL: A PROTOTYPE FOR GENERATIVE REPRESENTATION

The case study of Studious presents the hypermodel as a new form of modeling that facilitates diverse methodologies and content reaching beyond disciplinary boundaries and possibly creating new hybrids. As a historical reconstruction tool, the hypermodel is designed to support and enhance archival as well as interpretive research practices. The outcome of this project suggests future possibilities for implementing the hypermodel, not only for the purposes of reconstruction, but for reconceptualizing ways of engagement in and when building on complex sites. Therefore, an aspiration of this essay is to illustrate how the functional capacity of the hypermodel can be harnessed to reconsider the *concept* of modeling. The questions that the hypermodel addresses range beyond the unique example of Studious, or other historical sites, to offer new possibilities for digital representation of three-dimensional phenomena more generally. It allows for the autonomy both of representation and knowledge, that is, it presumes an irreducible difference between, say, a thing and its depiction—a map and the territory mapped, for example. Indeed, the hypermodel draws attention directly to the mediation of representations, including the ways that they are sometimes built up over time by

FIGURE 8.5. The "temporal layering" sequence captured as stills in two rows, read from left to right. The top half of the series starts with the volumetric rendition of the apse in the Byzantine era. In four snapshots, the rendering transitions from volumetric to diagrammatic, where the surfaces recede and the wireframe highlighting the geometry emerges. The fourth image represents the crux of the layering technique, in which the wireframe of the Ottoman-era apse massing is overlaid on top of the Byzantine version, revealing the differences in an analytical manner. The bottom row is a continuation of the sequence in reverse, where the diagrammatic lines recede, leaving room for the volumetric surfaces representing the Ottoman apse to emerge. Image by the author.

different actors under different conditions with sometimes divergent agendas, but without forgoing the possibilities of representation and reference altogether. The hypermodel advances the idea that complex material objects are to varying degrees palimpsests—in time, space, formation, alteration, use, and reception. Its modeling is therefore founded on a process of layering, with individual layers not only containing moments in time or particular spatial coordinates but also becoming accessible by different sensory means. In this way, the hypermodel promises a more comprehensive and manipulatable, and, ultimately, critical, means of understanding how our built environment has come to be the way it is.

NOTES

1 Cyril Mango, *Byzantine Architecture* (New York: H. N. Abrams, 1976), 62–70.
2 Feridun Dirimtekin, “Imrahor Camii,” in *TTotK. Belgeleri,* no. 37/316 (Istanbul: Turk Tarih Kurumu, 1973), 8–10.
3 Alexander Van Millingen, A. E. Henderson, W. S. George, and R. Traquair, *Byzantine Churches in Constantinople: Their History and Architecture* (London: Macmillan, 1912), 35–55.
4 Halil Inalcik, “The Policy of Mehmek II toward the Greek Population of Istanbul and the Byzantine Buildings of the City,” *Dumbarton Oaks Papers* 23/24 (1969–70): 229–49.
5 According to the Association for Conservation of Environmental, Cultural, and Historical Heritage of Istanbul (ISTED), http://isted.org.tr, which monitors the restoration and recommissioning of historical sites as mosques, since 2018, the project remains on the restoration list, but no work has begun as of 2020.
6 Vercihan Ziflioğlu, “Istanbul Monastery to Become Mosque,” *Hurriyet Daily News,* November 26, 2013, http://www.hurriyetdailynews.com/istanbul-monastery-to-become-mosque-58526.
7 Fatih Köse, “Imrahor Ilyas Bey Camii ve Osmanli Doneminde Gecirdigi Tamirler,” *Vakif Restorasyon Yilligi* 4 (2012): 32–39.
8 Mango, *Byzantine Architecture,* 64.
9 Van Millingen et al., *Byzantine Churches in Constantinople,* 53.
10 Mango, *Byzantine Architecture,* 62.
11 Sharon E. J. Gerstel and Julie Lauffenburger, eds., *A Lost Art Rediscovered: The Architectural Ceramics of Byzantium* (University Park: Pennsylvania State University Press, 2001), 27.
12 Esra Kudde and Zeynep Ahunbay, “Istanbul Imrahor Ilyas Bey Camii-Studios Bazilikasi Orta Bizans Dönemi Opus Sectile Döşemesinin Belgelenmesi ve Korunmasi Için Öneriler,” *Restorasyon Konservasyon Çalışmalari Dergisi* 17 (2014): 36–61.
13 Köse, “Imrahor Ilyas Bey,” 37.
14 Nicholas V. Artamonoff, Photographs of Istanbul and Turkey, 1935–1945, PH.BZ.010, Image Collections and Fieldwork Archives, Dumbarton Oaks, Trustees for Harvard University, Washington, D.C., http://images.doaks.org/artamonoff/collections/show/44.
15 Based on my research at the German Archeological Institute (DAI) in Istanbul.
16 NYU Institute of Fine Arts, “The Byzantine Churches of Istanbul,” https://ifa.nyu.edu/research/projects/byzantine/index.htm.
17 Dan Hicks and Mary C. Beaudry, eds., *The Oxford Handbook of Material Culture Studies* (New York: Oxford University Press, 2010), 14.
18 Wilhem Salzenberg, Paulus Silentiarius, and Carl Wilhelm Christian Kortüm,

Alt-christliche Baudenkmale von Constantinopel vom V. bis XII. Jahrhundert (Berlin: Ernst and Korn, 1854).

19 Byzantine and Christian Virtual Museum, http://www.ebyzantinemuseum.gr/.

20 Marie Stender, "Towards an Architectural Anthropology—What Architects Can Learn from Anthropology and Vice Versa," *Architectural Theory Review* 21, no. 1 (2017): 27–43.

21 Kudde, "Istanbul Imrahor Ilyas Bey Camii," 43.

22 For further investigation into historical reconstruction projects, read, e.g., Sheila Bonde, Alexis Coir, and Clark Maines, "Construction–Deconstruction–Reconstruction: The Digital Representation of Architectural Process at the Abbey of Notre-Dame d'Ourscamp," *Speculum* 92, no. S1 (2017): S288–320, or visit "Zamani Project" by University of Cape Town, http://zamaniproject.org/.

23 *Byzantium 1200*, http://www.byzantium1200.com/.

24 Robin Evans, *Translations from Drawing to Building and Other Essays* (Cambridge, Mass.: MIT Press, 1997), 154–86.

25 William R. Sherman and Alan B. Craig, *Understanding Virtual Reality: Interface, Application, and Design* (San Francisco: Morgan Kaufmann, 2003), 7.

26 Paul Emmons, "Drawn to Scale: The Imaginative Inhabitation of Architectural Drawings," in *From Models to Drawings: Imagination and Representation in Architecture,* ed. Marco Frascari, Jonathan Hale, and Bradley Starkey (New York: Routledge, 2007), 64.

27 Pierre Lévy, "Toward a Superlanguage," in *ISEA 94*, exhibition catalog (Helsinki: University of Art and Design, 1994).

28 Timothy Morton, *Hyperobjects: Philosophy and Ecology after the End of the World* (Minneapolis: University of Minnesota Press, 2013), 12.

29 The author gratefully acknowledges the generous support of the 2017 Arnold W. Brunner grant by the New York Center for Architecture: The American Institute of Architects, New York Chapter, and the Center for Architecture Foundation. This grant provided invaluable assistance by Chris McAdams, Julie Kress, Gaurav Guru, and Mike Primo.

30 Based on personal correspondence with the Directorate General of Foundations in 2016.

31 Ann Brower Stahl, "Material Histories," in Hicks and Beaudry, *Oxford Handbook of Material Culture Studies,* 6, 150–72.

Part IV

Doing

9

Modeling Maneuvers: Anatomical Illustration and the Practice of Touch

JULIET S. SPERLING

In the middle of the nineteenth century, cities including Philadelphia, New York, and London were in the throes of a public health renaissance, and portable anatomical models were bountiful. Crowds flocked to hear practitioners profess "physical education" or paid ten cents to enter a new anatomical museum—eager, in each case, to see the hidden interior of the body, pictured through an array of visual spectacles including paintings, wax casts, and papier-mâché manikins. Of this variety of techniques, structurally complex three-dimensional paper illustrations called "dissected plates" were the most widely used and easily accessed format.[1] Small in scale and designed by physicians rather than professional artists, they lack the jarring mimesis of eighteenth-century wax anatomical Venuses or the iconographic sophistication of easel paintings of the surgical theater, which have received far more scholarly attention.[2] Yet, unlike sculpted or painted representations of dissected bodies, which were almost always one-of-a-kind objects intended for public display, dissected plate illustrations were most often found within books, printed in runs ranging from the hundreds to the thousands and easily purchased at a local bookseller's shop, and were therefore experienced in a variety of settings and configurations. After watching an anatomical lecturer demonstrate operations with the aid of a dissected plate illustration, a curious audience member—perhaps an aspiring medical student, a retired surgeon, or an inquisitive member of the nonmedical public—could purchase his own copy, bound within an informative anatomical textbook, to practice the same maneuvers rehearsed before him in the lecture hall. Modest in scale, transportable, and accessible to many, using dissected plates transformed a viewer into a doer.

Dissected plates are models: objects that facilitate, through a particular combination of structure, scale, and image, movement between abstract or imagined ideas and physical actions. This definition of *model* is necessarily broad. As the diverse things and practices brought together in this volume show, attempting to identify a unified style or structure of models across time and space would be a futile task. But this lack of formal cohesion is precisely what makes models so ripe for analysis through a material culture lens, which (in its most basic articulation) contends that a society's foundational beliefs are often borne out in the physical shape of everyday things. Each model's particular configuration of form and representation—both what is included and what is left out—speaks to the values, ideologies, and priorities of its specific social and historic context.

Working toward a more granular understanding of modeling as a traversal of imagined and physical practice—of thinking and doing—therefore means approaching models as embedded within specific intellectual histories. This essay adds to this ongoing and collective project with a single case study that explores the forms and uses of paper anatomical models in nineteenth-century England and America as a lens onto the relationship between seeing and acting in a particular time and place and, as such, considers the interpretive potential of models as historical documents. It begins with a close visual and material analysis of the paper dissected plate models in George Spratt's 1833–50 publication *Obstetric Tables* before moving toward a broader social and historical contextualization, tracing how they revised and expanded upon the preceding illustrations of Jacques Maygrier's *Nouvelles démonstrations d'accouchemens* to meet growing interests in physically interactive learning.

Ultimately, this chapter aims to demonstrate how the translation of Maygrier's two-dimensional sequential pictures into Spratt's three-dimensional simulations opens up new ways of thinking about the affordances—and, just as importantly, the limits—of haptic interaction with models. Bridging public performance and private practice, specialized and lay audiences, and distant beholding and hands-on engagement, layered anatomical illustrations represent a missing element in the story of nineteenth-century constructions of knowledge: a period conviction in the inadequacy of eyesight alone. Unsatisfied with the hands-off experience of watching a lecturer operate models, audiences craved their own chance at tactile processing. Dissected plates, through their innovative sculptural construction, allowed users to do just that: penetrate the body's depths, visualize time-based organic processes, and rehearse the complex physical maneuvers of surgery with their very own eyes and hands. Contained within the broader media species of the illustrated scientific book, a genre that has long been discussed as a site in which text- and image-based forms of knowledge converge and coalesce, dissected plates offered a more sharply delimited space in which modeling's balance between imaginative processing and physical action could be tried and tested.

PAPER BODIES

George Spratt's *Obstetric Tables* was the most accessible and widely circulated dissected plate publication of its day.[3] Between 1833 and 1850, five editions and thousands of copies of the unusually formatted book, written and designed by an English botanist, printmaker, and male midwife named George Spratt, circulated around London and Philadelphia. Originally published in London and reissued there in multiple editions until 1846, Spratt's text was subsequently released in 1847 by Philadelphia publishing firm Wagner and McGuigan under the full title *Obstetric Tables: Comprising Graphic Illustrations, with Descriptions and Practical Remarks; Exhibiting on Dissected Plates Many Important Subjects in Midwifery.* Wagner and McGuigan maintained the book's core text but redrew its illustrations on stone down to the individual flap. Already familiar to American readers as a British import, the book gained greater popularity as a more affordable Philadelphia imprint. It was published again by Thomas, Cowperthwaite and Co. in 1848 and in 1850 by James A. Bill, both of Philadelphia, using Wagner and McGuigan's illustrations.

Between the quarto's plain cloth-bound covers, layers of hand-cut lithographic illustrations stack upon one another, forming movable images of a female body in various stages, angles, and predicaments. Each of these three-dimensional printed images comprises multiple layers of superimposed pictorial paper flaps. In some plates, sculptural accretions of paper slips mimic the swells and curves of pregnancy, while in others, razor-sharp edges of paper evoke the scalpel slices of surgery. Print-shop laborers sandwiched meticulously registered stacks of smaller tabbed illustrations between a base sheet and a top sheet, its surface slit to reveal the layers beneath. This structural technique had been used for flap anatomies or fugitive sheets (as earlier incarnations of dissected plates have been called) since the early seventeenth century.[4] Between the first London edition of 1833 and the first American edition of 1847, the book eventually grew from thirteen to twenty-one layered illustrations.[5]

Spratt did not invent the anatomical flap illustration. The genre was already well established in European print centers by the nineteenth century and had a presence in Japanese material culture as well.[6] But whereas European fugitive sheets made from the sixteenth century onward treated layers as indices of depth, the tactile images of *Obstetric Tables* added new dimensions to the centuries-old format. For three hundred years, anatomical flap prints had adhered to a conventional spatial metaphor of peeling back the body's skin, each successive layer representative of a deeper cut. In *Obstetric Tables,* rather than solely follow this excavation-style approach to visualizing dissection, some layered illustrations also worked to denote change over time, like a slow-moving flip book, while others detailed the intricacies of physical maneuvers—application of pressure, grasping a tool—through

FIGURE 9.1. *Left,* lithograph by Wagner and McGuigan after George Spratt in *Obstetric Tables: Comprising Graphic Illustrations, with Descriptions and Practical Remarks, Exhibiting on Dissected Plates Many Important Subjects in Midwifery* (Philadelphia: Wagner and McGuigan, 1847). Kislak Center for Special Collections, University of Pennsylvania. Photograph by the author. *Right,* lithograph by Dewasme after Antoine Chazal in J. P. Maygrier, *Nouvelles demonstrations d'accouchemens* (Bruxelles: A. Wahlen, 1825–27). Wellcome Library, London, 36048/C. Public domain.

the same stacked format. Disembodied hands of surgeons and attendants, scattered across these flaps in ghostly black-and-white outlines that contrast markedly with the colored solidity of the depicted bodies, primed the way for the reader's own points of contact with the page. Using kinetic layers to such ends was strikingly new: Spratt's tactile images did not simply analogize and mimic excavations into the three-dimensional corpse. Beyond modeling matter fixed in place, they provided the user with a means of practicing maneuvers and time-based actions, a nascent interactive technology of simulation keyed to an array of medical scenarios.

Multiple types of models are contained within the covers of *Obstetric Tables,* and each offers its own formula for linking concept with physical action. The fifteen kinetic plates interspersed with the text fall into three categories. At the most basic level, stacked flaps index spatial relationships, useful for visualizing dissection or internal examination. A second category deals with temporal problems, for instance, tracking changes in the body over the nine months of pregnancy. The third category illustrates spatiotemporal scenarios, specifically, manual operations that require a set of surgical procedures. Period reviewers made note of these three distinct groups and praised this functional diversity as one of the book's greatest strengths.[7]

Category 1, in which paper layers index actual bodily space, is neatly demonstrated by the first movable illustration in *Obstetric Tables*: Table III, a straightforward view of female genitalia. As I discuss in depth later in this essay, the design for Table III, like many of Spratt's illustrations, is drawn directly from Jacques Pierre Maygrier's earlier obstetrical atlas *Nouvelles démonstrations d'accouchemens*

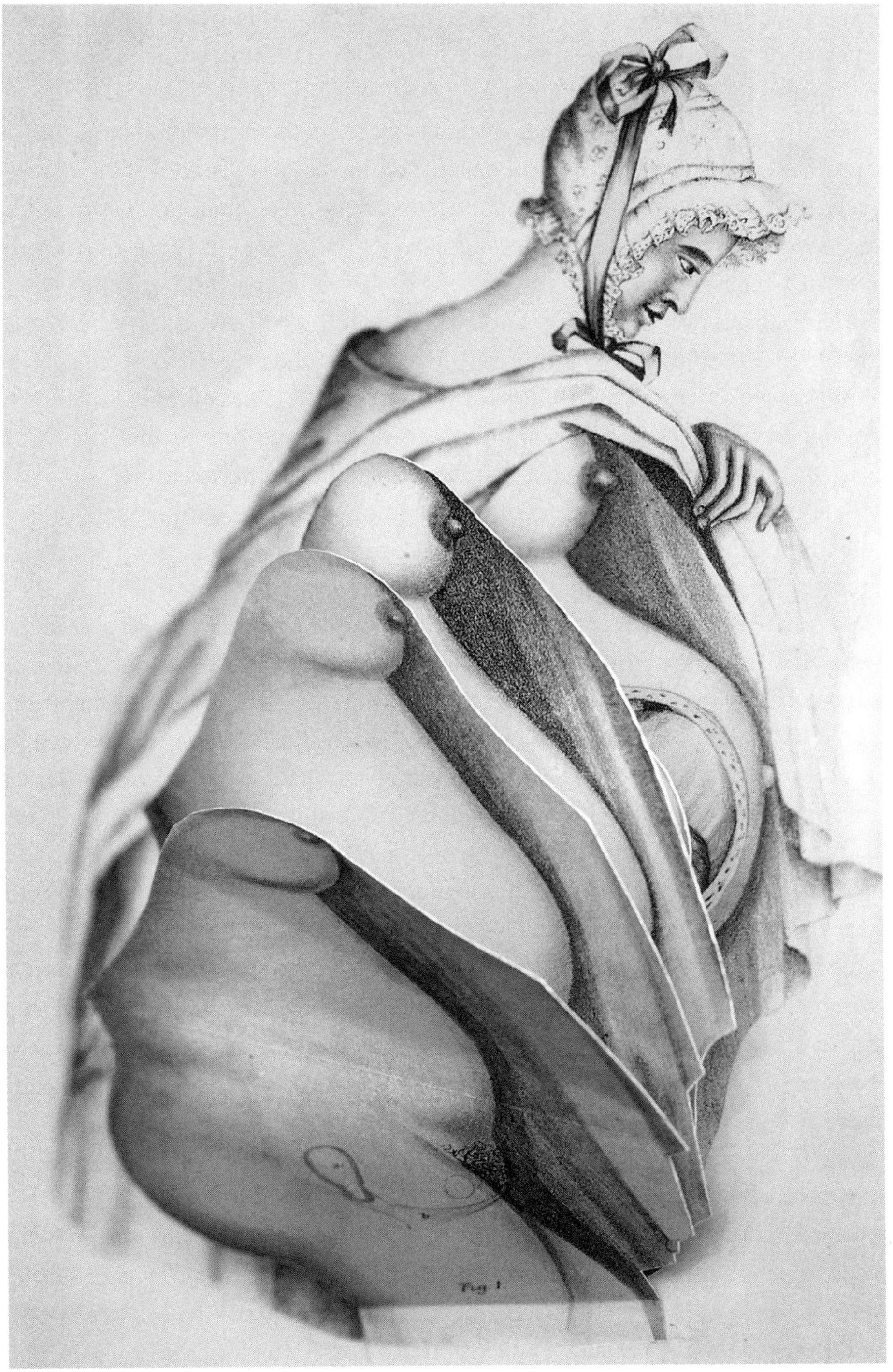

FIGURE 9.2. Changes in the body over nine months of pregnancy depicted through layered flaps. George Spratt, Table IV in vol. 1 of *Obstetric Tables* (London: J. Churchill, 1835). Courtesy of the National Library of Medicine.

and then enhanced by the novel mechanical format (Figure 9.1).[8] Thus, unlike Maygrier's fixed two-dimensional view, the flaps in Spratt's image correspond directly to the contours and depths of female anatomy. At the text's direction, the reader can "separate" the outermost flaps—here the labia majora—and perform a simulated internal gynecological exam. Similarly, Table XI allows the reader to dissect the uterus at full term, separating its numerous membranes into distinct paper layers to best compare differences in texture, color, and translucency. In a time when accessing cadavers for authorized dissection was still only newly legal, practicing dissection on a pregnant woman at full term would have been a rare experience (not to mention a time-sensitive one, in the prerefrigeration age), and this illustration is a prime example of how paper technologies of modeling helpfully sidestepped obstacles of access and preservation, though the replacement would always be necessarily incomplete and inadequate in comparison.

One of the most frequently reproduced images from this still widely collected and exhibited book neatly exemplifies the second temporal category of dissected plates (Figure 9.2). Appearing in editions printed from 1835 onward, Table IV, "On the Signs of Pregnancy and the Development of the Uterus," depicts a woman in profile view from head to thigh against a blank background. She is nude, save for a cloak loosely draped over her shoulders to signal her modesty and a frilly bonnet secured to her head by a bright blue ribbon. Cheeks flushed, she demurely diverts her gaze away from the reader, who is about to manually enter her body. Between the edge of the cloak and the top of her breast, a slight fluctuation in the paper's thickness alerts the user to an opening between the first layer and the base of the page. As each successive flap lifts, the woman's body propels through nine months of pregnancy, belly swelling from sheet to sheet, nipples elongating and darkening in color—pigment likely hand-applied by female printshop laborers. Opening the innermost flap uncovers a small, curled fetus. Overall, four small paper flaps encapsulate the woman's transition from "virgin female" to "the female at full period of gestation."[9] As the reader pages onward, the model's body becomes increasingly fragmented and depersonalized, the object of a Foucauldian medical gaze.[10]

The remaining tables fall into the third and most complex category: images that represent operations, in every sense of the word. These dissected plates allow readers to rehearse gestures and physical maneuvers in three dimensions, demonstrating surgical tools like obstetrical forceps in action. In each of these images, users' fingers are guided through a simulated tactile experience: a physical examination of the patient, a partial-birth abortion, a C-section surgery. Paper flaps compress spatial and temporal concerns into a single material unit in these illustrations, highlighting both the promise and pitfalls of tactile interaction as a method of learning from visual surrogates.

FROM PICTURE TO MODEL

To period readers, the publication's allure was in these elaborate, touch-activated movable pictures, rather than the written descriptions of procedures familiar from the popular eighteenth-century anatomical atlases that still defined the medical literature of Spratt's day. One reviewer noted the established yet unimaginative nature of the text, writing that "the descriptions and directions are brief, explicit, and generally sound—in fact, the doctrines are essentially those of [Thomas] Denman and [William] Smellie."[11] The paper models, conversely, were the source of more excitement. Though this chapter's primary aim is examining why the paper models elicited such an enthusiastic response, it is important first to establish exactly who was (and was not) responding, how they arrived at the object, and how the range of options available to them shaped the expectations they brought to viewing.

Bound within books, dissected plates were positioned at the nexus of overlapping circuits of visual communication delimited by race, gender, and class. At the time of the *Obstetric Tables*' publication, middle- and upper-class white men could choose from an array of experiences to learn about the workings of the body: they might attend a public lecture, view a wax-fleshed Venus sculpture undergo dissection in an anatomical museum, or enroll in medical school to take part in dissections themselves. Some white women also had opportunities to learn about their bodies, albeit from a more limited set of choices. They could participate in a range of popular anatomical performances, lectures, and courses from the 1830s onward (in some cases, segregated by gender), though they did not gain access to the official instruction of medical schools until the 1850s. African Americans, Native Americans, and other marginalized groups had drastically fewer options for accessing scientific information, a bitter irony given the history of using Black and Indigenous bodies to produce and test that knowledge. These populations had long been associated with anatomical study in violent and dehumanizing ways, as most of the dead bodies plundered for dissection in the years before the Anatomy Acts of the 1830s were those of African Americans, Native Americans, or members of marginalized immigrant groups. Although the establishment of medical schools like Philadelphia's Woman's Medical College, which admitted a diverse student body from its founding in 1850, opened some doors to Black women interested in gaining medical knowledge, many continued to shy away from the world of anatomy because of its nefarious history.[12]

For those privileged to have options, each venue was endowed with its own benefits and drawbacks considered by the viewer. For instance, public lectures often prioritized entertainment over scientific fact, while popular anatomical museums presented a skewed view of subjects usually "excluded from bourgeois identity"—

the diseased, erotic, or otherwise "freakish."[13] Those with the freedom to choose from a full menu of scientific entry points turned to paper models knowing that the format held both advantages and disadvantages. And to be sure, reviewers of Spratt's *Obstetric Tables* acknowledged and debated the strengths and weaknesses of the paper model for particular readerships. Many believed that it was an object best suited for medical students and practicing physicians—that the unusual visual aid would help young doctors learn and would jog the memories of the older set growing rusty in their late careers.[14] Some went so far as to claim that anatomical inaccuracies and exaggerations of scale in the illustrations were a strength, a way of highlighting important body parts that would be otherwise difficult to notice in the messy viscera of dissection.[15] Still others disagreed, insisting that while the book held aesthetic appeal for the scientist or "exacting bibliomaniac,"[16] its miniaturized paper surgeries could never effectively simulate the textures, sounds, and adrenaline of the anatomical theater.[17] From these converging factors, a picture of the book's readership emerges: white, primarily male, and most likely professionally connected with medicine.

Given, then, the medical expertise readers brought to *Obstetric Tables*, their zeal for its models despite clear limitations is perhaps surprising. Far from skeptical dismissals, these final comments are found in reviews that are comprehensively praiseworthy—an initially puzzling disjuncture that forces us to reconsider the criteria for a successful model. If the 3D paper models of *Obstetric Tables* could admittedly not truly teach readers to perform surgery, then what knowledge *did* they impart? Looking to Spratt's source material sheds light on this question. As I have mentioned, many of the illustrations in *Obstetric Tables* were first found in two-dimensional form in *Midwifery Illustrated* (1822), an obstetrical atlas that elevates hand movements and gestures as markers of knowledge and sophistication on their own terms, both in and out of medical practice. A comparison of the two publications reveals how differences in format, physical structure, and relationship between image and material support can radically change the meanings and uses of nearly identical iconography—in other words, a case study for how an image can become a model.

French physician Jacques-Pierre Maygrier, the author of Spratt's visual source material, authored several publications focused on visualizing the manual and physical techniques of obstetrics through illustration.[18] *Nouvelles demonstrations d'accouchemens* (retitled *Midwifery Illustrated* in translation) was his final and most ambitious. Featuring lavish illustrations designed by Antoine Chazal and engraved on copper by Charles Aimé Forestier and François-Louis Couché, *Nouvelles demonstrations* led the charge in a new anatomical publishing trend that prioritized the visual. Maygrier's text was first published as a luxurious multivolume folio in Paris

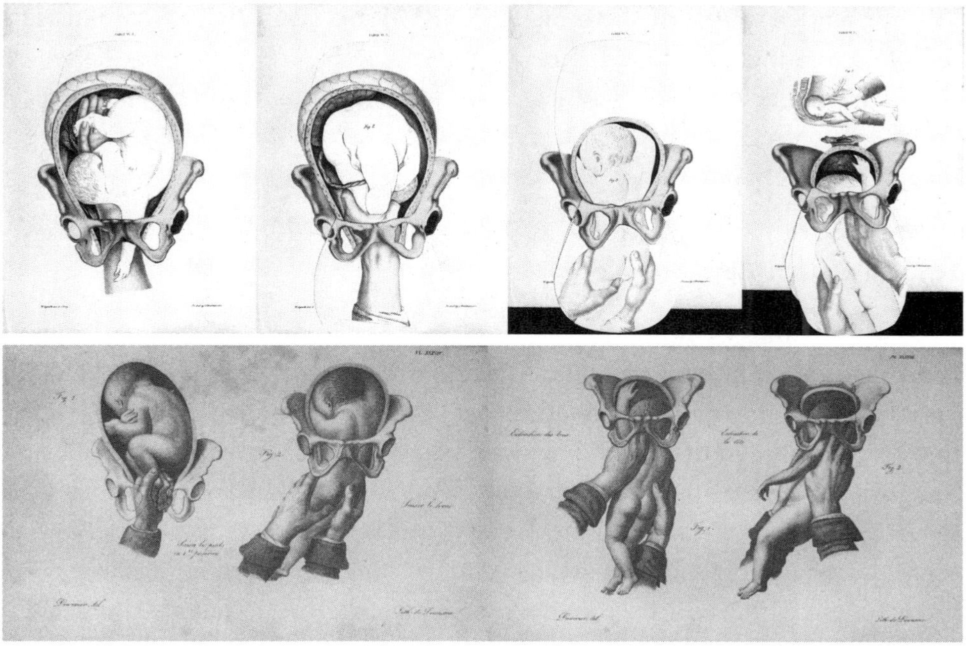

FIGURE 9.3. Shown here in their individual layers, the images above compose a single dissected plate with double-sided flaps that simulate the motion of turning a fetus when opened in succession. In contrast, Maygrier's book diagrammed the procedure in sequential images across several pages. *Above,* George Spratt, Table VI B in vol. 2 of *Obstetric Tables* (London: J. Churchill, 1835). University of Glasgow Library. Public domain. *Below,* lithographs by Dewasme after Antoine Chazal in J. P. Maygrier, *Nouvelles démonstrations d'accouchemens* (Bruxelles: A. Wahlen, 1825–27), Plates XXXVII and XXXVIII. Wellcome Library, London, 36048/C. Public domain.

in 1822, and publishers commissioned scaled-down lithographic copies of Chazal's illustrations when the book subsequently was reprinted in a smaller, cheaper octavo format across Europe, the United States, and Mexico.[19] The images would appear in other publications well into the late nineteenth century, often orphaned from their initial textual referents.[20]

Placed side by side, *Nouvelles demonstrations* and *Obstetric Tables* appear as two potential solutions to the same problem: how to depict the complicated time-space confluences of pregnancy within the limited physical structure of the book. As we now know, Spratt chose to bend the book's basic architecture to the very boundaries of what it could support, coaxing three dimensions out of two. Yet before Spratt's efforts, Maygrier and Chazal turned to a time-tested alternative solution in the sequential image. Take, for example, Table VI B's tumbling fetus. In the Maygrier plates, it is as if Spratt's stacked layers have been peeled apart and pasted next to one another in a straight line across the page. The sources for Spratt's Table VI B are four figures split across two plates, XXXVII and XXXVIII, which illustrate the same rotating, pulling, and grasping motions (Figure 9.3). In the flattened, sequential redistribution of these stages, Spratt's images undergo an archaeological regression: one can see how models that invite a user to *do* derived from images that ask a viewer to *look*.

TEACHING TOUCH

Despite these images' primarily visual address, Maygrier was, like Spratt, chiefly concerned with conveying tactile knowledge through the book as a whole. Whereas Spratt's kinetic illustrations offered readers a chance to put physical knowledge into practice, readers of the Maygrier were encouraged to move between language, imagination, and (if they were practitioners) the real scenario. In the text, he recognizes that operating with instruments was something that must be learned through repeated practice; touch needed to be taught. Calipers, for instance, were prone to "errors into which we may consequently fall from a want of experience."[21] The pelvimeter, too, was "ingenious in its combinations [but] presents numerous inconveniences, which have caused it in a great measure to be abandoned: these are, the difficulty and the danger attending its introduction."[22] In the most dire of scenarios, Maygrier recommended assessing the situation "by touching: by introducing the index finger of one hand into the vagina, the practitioner can acquire all the knowledge relative to the examination he proposes to make."[23] In some shocking plates, Chazal's illustrations grotesquely reduce the female body to skinned pelts tacked to the page by their edges, as if to further invite physical interaction unobstructed by the movements and volumes of an embodied human being (Figure 9.4). Yet recommending that the student relinquish book study in favor of hands-on instruction alone would be poor business practice for an author, whose livelihood depended on a strong rationale for his work's purchase. Therefore Maygrier, and later Spratt, emphasized the illustration as a site of an equally important companion to intellectual and physical practice. Even if it were possible to "acquire all the knowledge" with an index finger alone, the book served as a necessary handbook for what knowledge to seek and how to seek it.

Much of the rhetoric of doing—to maneuver, to manipulate, to manufacture—derives from the Latin *manus,* or "hand." In both Spratt and Maygrier, an emphasis on illustrated hands at once reinforces the period interest in manual interaction as a favored mode of absorbing information and calls its actual effectiveness as a tool for practicing real surgical movements into question. Consider Table VIII, Spratt's illustration of the proper use of forceps, a process more easily understood experientially than didactically (Figure 9.5). Each layer emphasizes the precise

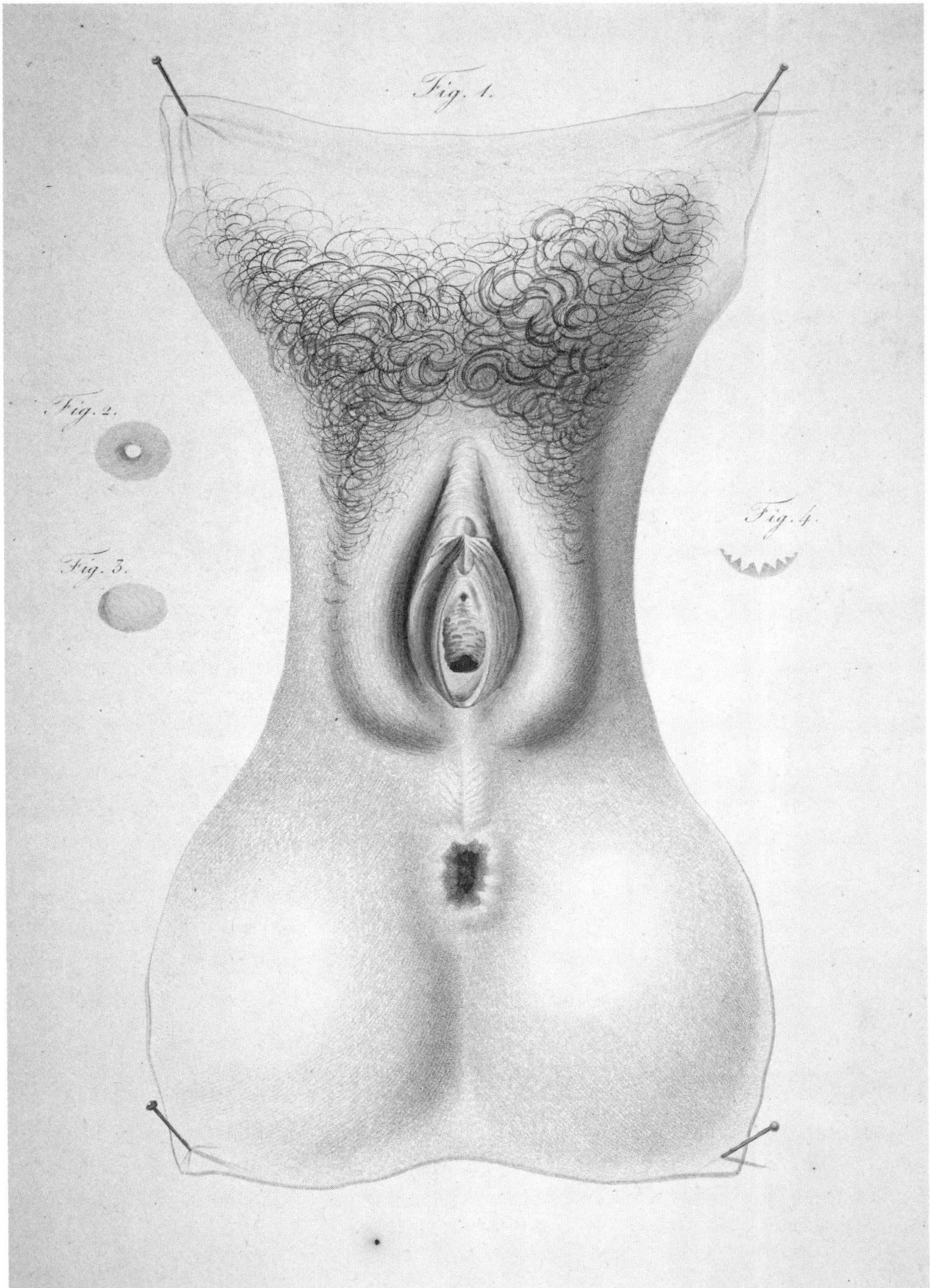

FIGURE 9.4. Chazal represents the female body as if it were an animal or insect specimen, tacked up with pins for meticulous examination. Hand-colored engraving by Forestier after Antoine Chazal in J. P. Maygrier, *Nouvelles démonstrations d'accouchemens* (Paris: Béchet, 1822–27). Wellcome Library, London, 36047/D. Public domain.

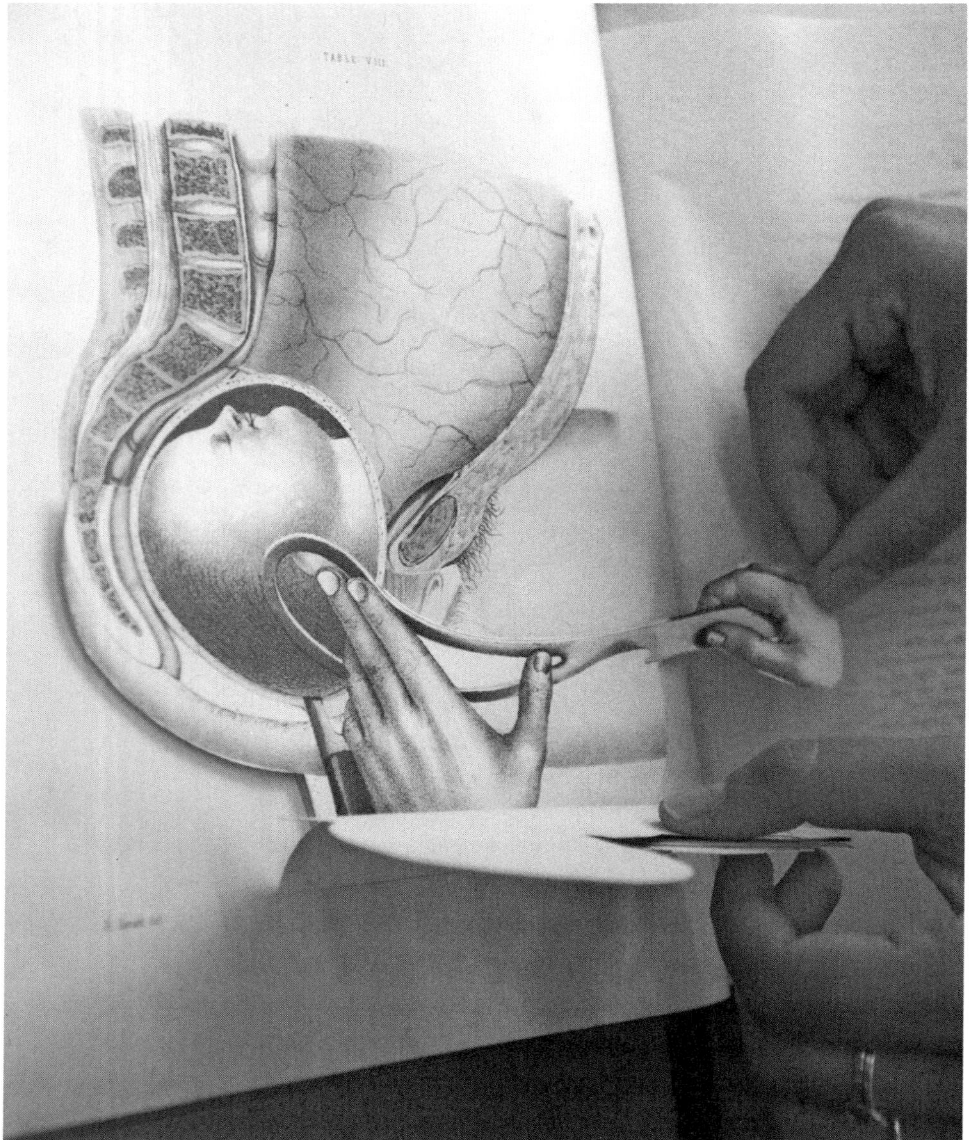

FIGURE 9.5. Pop-out hands point the way through a procedure in Table VIII of Spratt, *Obstetric Tables* (London: J. Churchill, 1835). Kislak Center for Special Collections, Rare Books, and Manuscripts, University of Pennsylvania. RG652.S7670 1847. Photograph by the author.

slant at which the tool must enter the body, stabilized by an exacting arrangement of fingers. A complex diagram, its successive flaps visually and structurally map out sensory qualities that would escape words for anyone without extensive training in the proper use of instruments: directionality, force, pressure, and texture. To translate sensory cues into the paper simulation, Spratt used illustrated hands,

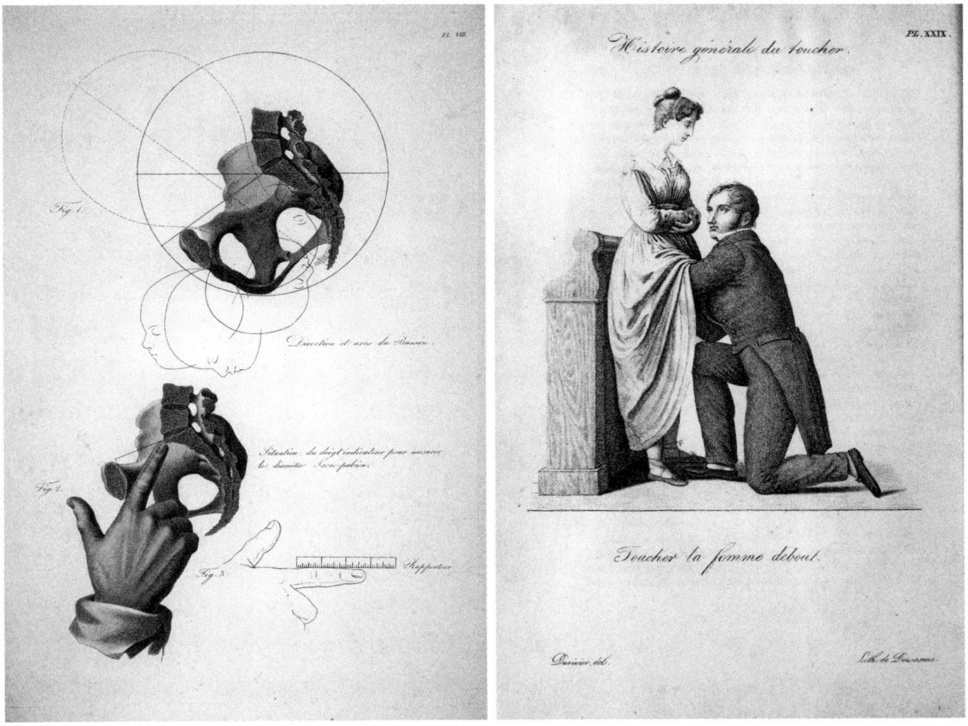

FIGURE 9.6. *Left,* readers of *Nouvelles démonstrations d'accouchemens* are shown how to use their hands as tools of medicine. Engraving by Forestier after Antoine Chazal in J. P. Maygrier, *Nouvelles démonstrations d'accouchemens* (Paris: Béchet, 1822–27). Wellcome Library, London, 36047/D. *Right,* lithograph by Dewasme after Antoine Chazal in J. P. Maygrier, *Nouvelles démonstrations d'accouchemens.* Wellcome Library, London, 36048/C.

which literally pop out from the page, to connect the user's own body with that of the pictorial surgeon. Cupping the forceps from beneath is the surgeon's second hand, again truncated just below the palm. Throughout the course of the illustration's five flaps, the outlined hands subtly change position. A thumb circles the forceps' handle to re-angle the tool, as the other hand's fingers slip protectively beneath the fetus's head. Eventually, the flaps burrow too deeply inside the body for hands to fit, and the fetus's head is alone with the metal blades of the tool. To proceed with the scenario, the reader must join his own hand with the illustrated paper appendage, which floats above the page, inviting precisely this symbolically charged interaction between user and paper surface.

Nowhere is this emphasis on acquiring tactile knowledge more evident than in Table VI B, which visualizes the process of manually rotating a breech fetus to achieve a safer presentation of the head (Figure 9.3). Version, or the operation of turning, is a primarily manual process in which the midwife is deprived of visual clues. Instead, she or he must know by sense of touch precisely where to grasp for the fetus's limbs, how much pressure to apply, and what direction to rotate. Though the table comprises only three flaps, its design is visually complex and unique within the book as a lone experiment with illusory motion, the style of movement most commonly recognized as protocinematic.[24] Three slips of paper printed on both sides show four successive stages in the operation of turning. When tasked with illustrating a time-based operation like version (external cephalic version, the medical terminology for turning a breech baby in the womb), double-sided images work to eliminate any gaps in the fluid sequence of stages. In the top flap, the midwife carefully grasps the breech baby's upended feet. Opening the flap downward reveals, on its verso, the midwife's firm hand beginning to pull the fetus by its feet toward the opening of the cervix. By rhyming the physical movement of the flap with the trajectory of the fetus, the reader appears to convincingly tug the baby downward on the page as she operates the illustration. Spratt's tactile image gives sight to an otherwise blind process, with the final goal of teaching the operator how to see with her hands.

In Maygrier's text, several plates of hands and fingers in precisely diagrammed poses slow down the blur of live study to an analyzable crawl (Figure 9.6). One illustration shows the reader how to turn the index finger into a mathematical instrument capable of calculating the "true measure of the antero-posterior or sacro-pubic diameter" if inserted at precisely the right angle.[25] Another image provides an important sensory footnote on social decency, depicting the doctor politely averting his gaze as he conducts his digital examination. Not only was touch-based investigation superior to (and more practical than) visual examination; in this case, it was also a mark of a good bedside manner.

In elevating the tool and the maneuver, Maygrier and Spratt consciously waded into lively debates about a new controversial figure in the medical field: the so-called man-midwife. Childbirth had long been the purview of female practitioners, but in the late seventeenth century, men began to encroach on that space—though not without significant public and professional pushback. Given this context, it is no accident that Spratt's illustrations foreground hands as the physical means of entry into manipulating his paper machines. The hand, and particularly trained, nimble movements of the fingers, took on symbolically charged meaning in the context of debates about the gendered professionalization of the field of women's health from the mid-eighteenth century onward. British male midwives (who would not be called obstetricians until about 1820) pushed for recognition as a professional, bourgeois class among their fellow physicians during the late eighteenth to early nineteenth centuries.[26] Male midwives occupied an uneasy position in this professional medical hierarchy: they competed with female midwives for business, while public opinion variously figured them as fashionable dandies, quacks, and

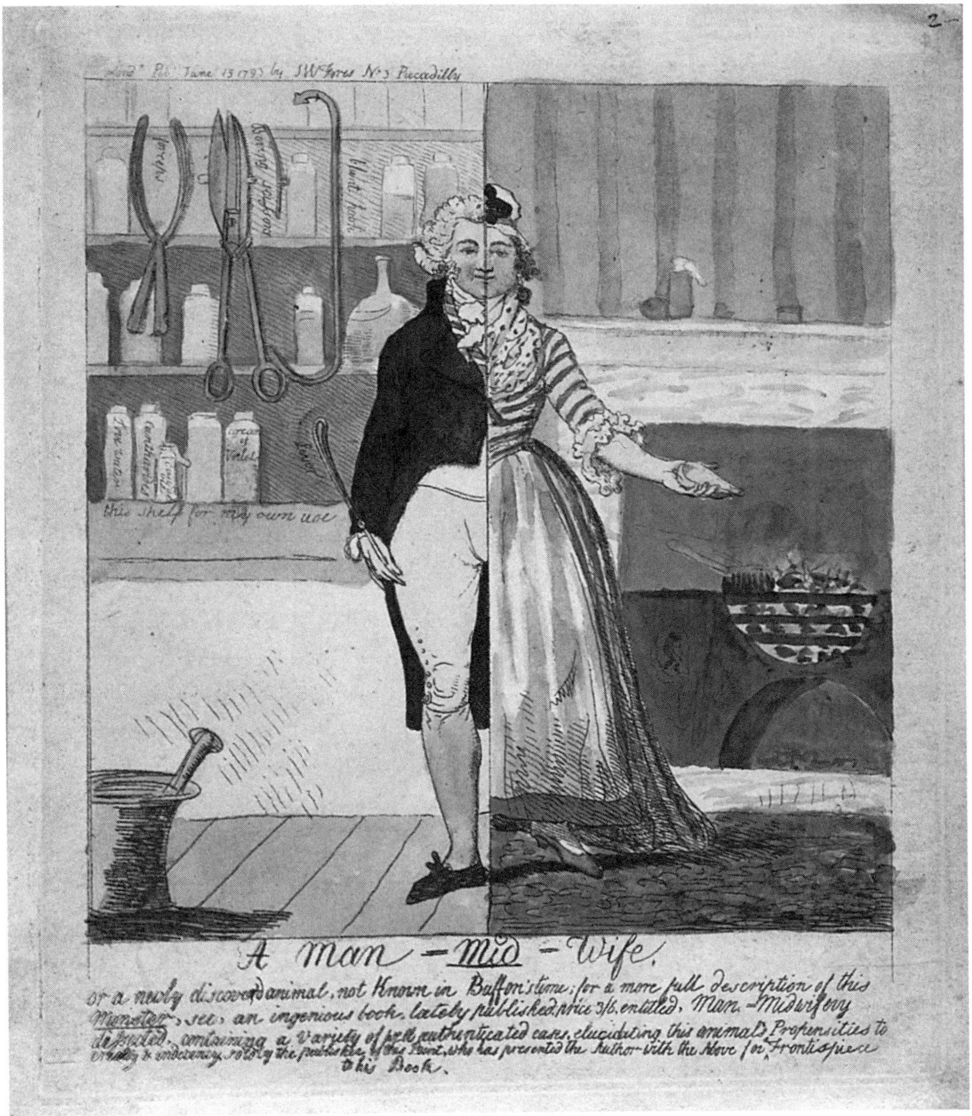

FIGURE 9.7. One of many images skewering the controversial figure of the male midwife. Cruikshank's caricature aligns obstetrical instruments with conventional masculinity. Isaac Cruikshank, a "man-midwife" (male obstetrician) represented by a figure divided in half, one half representing a man and the other a woman, June 15, 1793. Hand-colored etching. Wellcome Collection no. L0018481. Public domain.

gender-bending freaks.[27] Isaac Cruikshank's *Caricature of a "Man-midwife" Represented by a Figure Divided in Half* summarizes those denunciations, implying, with disgust, that men who assisted with labor must transgress accepted boundaries of gender and sexuality (Figure 9.7). At left, outsized forceps define (or perhaps disguise) the male midwife's genteel professional half, dangling prominently over a bottle of "love water" that hints at nefarious intentions. As Cruikshank's caricature exaggerates, proficiency with tools and physical maneuvers became a powerful weapon of self-fashioning for this new professional class—a means of signaling education in anatomical study and experience in all types of medical situations, while most of all preserving conventional masculinity.[28]

Male midwives' professionalization strategy depended on demonstrating a need for their services dire enough to excuse breaking social taboos about touching married women's bodies.[29] By presenting themselves and their physical toolkits as the only thing standing between life and death, they successfully redefined their profession as an indispensable necessity.[30] Public image was critical to this performance of competence: well-respected practitioners of the day publicly eschewed any visible signs of eccentricity that might call a male midwife's respectability into question. "I think it unnecessary, and would wish that any thing which adds to the mystery or peculiarities of the profession should by all means be avoided," recommended one doctor in 1833.[31] The male midwife persona was thus a cultivated performance of education, physical skill, and social grace, meant to appeal to clients of means faced with the choice between a traditional midwife and a self-fashioned "surgeon-accoucheur." Seen in this light, Spratt's illustrations function as low-stakes proxies for medically *and* socially high-stakes scenarios. They convey via tactile interaction sensory qualities that would otherwise escape words, especially for those without hands-on training in the proper use of instruments, instilling confidence in doctor and patient alike.

Situating Spratt's book within the context of *Nouvelles Demonstrations*'s earlier emphasis on the manual details of doing obstetrics shows that by the time of its publication, the role of touch in the medical investigation of women's bodies was so firmly established that it had acquired a codified visual shorthand. Pointing, grasping, active fingers punctuate its illustrated leaves as a reminder of the equal importance of physical and intellectual access in the cultivation of medical knowledge. Spratt's design capitalized on this newly established visual style and enhanced it with a kinetic element, thereby destabilizing the mental–tactile balance that earlier texts had presupposed and placing physical investigation in the position of primacy. As many material text scholars have noted, books are inherently tactile containers of information that must be physically handled to do their work.[32] The reader's body becomes the conduit for the flow of knowledge, setting a personalized pace with each turn of a page. Much like the indices or manicules that have peppered pages of manuscript and print since at least the twelfth century, these pointing hands served as a reminder of the charged bond between the reader's body and the materiality of the book's surface.[33]

Obstetric Tables emerged from an established scientific book culture that took touch seriously as an investigative methodology. Far from the novelty that previous studies have suggested, Spratt's book instead used its complex trifold program of moving images as a perceptive and shrewd attempt to outstrip the competition, throwing its full weight behind touch-based instruction. Spratt, after all, was both author and designer of his text, unlike Maygrier. Reading the book as the author and publishers intended pushed the reader to penetrate ever deeper into the secrets of the medical female body, each procedure or operation more complicated than the last. Most of all, the book's careful outlining of three distinct forms of movement positioned tactile investigation as something far from innate but instead specialized and practiced. Like learning to read or to speak a foreign language, accessing information with the hands was thought of as a skill acquired with the assistance of education and book study. Although this prizing of tactile knowledge was born from the medical community's attempts to underscore their professional indispensability in a moment of changing technologies in the field, it resonated with the priorities of a broader grappling with a shifting role of touch in spectatorship.

PRACTICING POWER

Tracing how *Obstetric Tables*'s three-dimensional simulations borrowed from and revised Maygrier's two-dimensional images allows for a fuller understanding of what interaction and tactility could—and could not—add to the picture. As period critics were wont to point out, scaled-down paper sculptures could never truly mimic the viscera and adrenaline of the operating room; but the success of the *Obstetric Tables* in spite of its clear material and mimetic limitations reveals that touch—and modeling technologies able to simulate manual experience—had symbolic, cultural stakes separate from the actual practice of medicine. Maygrier's engravings, which diagrammed a doctor's hand movements to a precise mathematical degree, speak to a period definition of medical professionalism through mastery of maneuvers and tools. Spratt's images build upon that baseline to seize upon the practice of the maneuver itself, revealing how the choreography of doing—the pulling, pushing, and grasping tactics of obstetrics—took on increased importance in a nineteenth-century culture that equated tactile knowledge and physical fluency with scholarly expertise and social sophistication. But the dissected plates of *Obstetric Tables* gave the reader a chance to do more than simply model maneuvers: they implicitly tie the performance of expertise to gendered fantasies of power and dominance.

Within gender and sexuality studies, there has been a long-standing awareness of how the female body has borne the weight of articulating ideas about matter, identity, and who is counted as human, and of science as a place where gender boundaries are articulated and reified. As the work of feminist scholars including

Elizabeth Bronfen, Luce Irigaray, Ludmila Jordanova, and many others makes clear, it is significant that the vast majority of dissected anatomical models (let alone dissected plates and flap anatomies), from the early modern period to the present, slice apart the female form.[34] It is crucial to note that *how* that slicing is pictured and structured, and what it implies about period configurations of knowledge, desire, and control, demands far more analysis than the scope of this essay allows. For instance, freezing a naturalistic, ecstatic female form in wax preserves space for the fantasy of the depicted body's willing response. Maintaining the complete contours of the body, particularly its fleshy surface, positions the corpse as a mere sleeping beauty, poised to awaken at the user's beck and call—a dynamic mirrored in Spratt's illustration of the blushing, bonneted woman. As period sources narrated (in ways that echo Pygmalion-esque reception of nude sculpture ongoing in the same moment), the pale corpse might spring to life at the graze of a viewer's fingertips.[35] Conversely, Maygrier's skinned pelts occupy the opposite end of this spectrum, comparing a conscious human to an animal whose ability to think or feel is voided along with the contents of its body. Spratt's flattened layers exist somewhere between these two poles. Constantly oscillating between flatness and three-dimensionality, the book's paper machines imagine women as a series of nested and evacuated shells, devoid of any interiority that would obstruct practice, all the while preserving just enough indication of humanity to remind its users that decorum required practice, too. In its invitation to manipulate and manhandle, the paper models of the *Obstetric Tables* provide a space for its user not simply to rehearse professional sophistication, but to do those things on the playground of women's bodies.

Let us return once again to the central tension presented by Spratt's dissected plates: though ineffective in simulating the knowledge necessary to truly practice obstetrics, the models were nonetheless praised, valued, and enthusiastically used. Despite their acknowledged constrained functional use value in the operating room, the books continued to sell and paved the way for countless other examples of the format throughout the nineteenth century and into the twentieth. Dissected plates were far from a gimmick—they had real cultural traction. And when considered in light of the gendered relationship between empowered viewer and dehumanized subject set up within, the popularity of practically inadequate nineteenth-century paper medical models takes on new layers of meaning. In addition to the perceived value of their mechanisms for modeling the professional performance of tactile knowledge, the paper surfaces supported more illicit desires regarding the control of female anatomy. Evidence for the coexistence of practical and erotic potential is found not in material traces but in absences. For example, I have encountered multiple copies in which the plate illustrating the "Female Organs of Generation" has been carefully excised from the book's spine. Divorced from the technical language of the text, the plate is freed to model whatever the reader desires.

Ultimately, models, as stated at the outset of this essay, are things that facilitate travel between imagination and practice. As the case of Spratt's *Obstetric Tables* shows, those traversals must be understood as fundamentally multidirectional. Manipulating layered paper dissections sent a nineteenth-century reader on a journey from a fuzzy mental concept of surgical operations to a deeply simplified physical rehearsal of those once-imagined movements, that is, from abstract idea to tactile action. But that reader could just as easily make a return trip to a state of imagination. "Doing" the paper surgeries hidden between the covers of *Obstetric Tables* sparked a chain reaction of imaginative associations: professional success, cultural sophistication, even the desire to touch a body without scientific motives. As material waystations between thought and activity, models invite repeated exploration.

NOTES

1 Nineteenth-century readers were familiar with the terminology *dissected plates* well before Spratt's publication appeared, e.g., "Mr. Tuson's Dissected Plates of Hernia," *Medical Quarterly Review* 2 (1834): 145–46.

2 E.g., see the large body of literature on Thomas Eakins's *The Gross Clinic* and other images of surgery or feminist theoretical studies of the anatomical Venus. Elizabeth Johns, *Thomas Eakins: The Heroism of Modern Life* (Princeton, N.J.: Princeton University Press, 1991); Elisabeth Bronfen, *Over Her Dead Body: Death, Femininity, and the Aesthetic* (New York: Routledge, 1992); Joanna Ebenstein, "Ode to an Anatomical Venus," *WSQ: Women's Studies Quarterly* 40, no. 3 (2013): 346–52.

3 James A. Secord, "Scrapbook Science: Composite Characters in Late Georgian England," in *Figuring It Out: Science, Gender, and Visual Culture* (Hanover, N.H.: Dartmouth College Press/University Press of New England, 2006), 164–91; Marcia Nichols, "Midwifery Demystified," in *Hidden Treasure: The National Library of Medicine*, ed. Michael Sappol (New York: Blast Books, 2012), 50–51.

4 Meg Brown, "Flip, Flap, and Crack: The Conservation and Exhibition of 400+ Years of Flap Anatomies," *Book and Paper Group Annual* 32 (2013): 6–14.

5 By first English edition, I refer to both volume 1 of the 1833 issue and the supplement, which is bound separately but was likely published very shortly after the first volume was released. Volume 1 of the 1833 printing includes nine layered illustrations and three regular illustrations; the supplement includes four layered illustrations and two regular illustrations.

6 Suzanne Kathleen Karr Schmidt, *Interactive and Sculptural Printmaking in*

the Renaissance, Brill's Studies in Intellectual History 270 (Boston: Brill, 2017); Suzanne Karr Schmidt and Kimberly Nichols, *Altered and Adorned: Using Renaissance Prints in Daily Life* (Chicago: Art Institute of Chicago, 2011); "An Edo-Period Flap Anatomy Manuscript," Thinking 3D, https://www.thinking3d.ac.uk/JapaneseManuscript/.

7 "Bibliographical Notices," *American Journal of Medical Sciences* 15 (1848); *The Western Lancet and Hospital Reporter* (Cincinnati, Ohio: Robinson and Jones, 1848), 165.

8 John Leonard Thornton and Carole Reeves, *Medical Book Illustration: A Short History* (Cambridge: Oleander, 1983), 108.

9 These descriptions come from the plate's accompanying text, "Description of the plate IV." This edition is unpaginated. Spratt, *Obstetric Tables.*

10 For more on the construction of the sexed and erotic anatomical female body, see L. J. Jordanova, "Medical Images of the Female Body," in *Sexual Visions: Images of Gender in Science and Medicine between the Eighteenth and Twentieth Centuries*, 134–59 (Madison: University of Wisconsin Press, 1989). Foucault's concept of the "medical gaze" is developed throughout Michel Foucault, *The Birth of the Clinic* (London: Routledge, 2003), 15.

11 According to one period review of Spratt's book, "the descriptions and directions are brief, explicit, and generally sound—in fact, the doctrines are essentially those of Denman and Smellie." Dr. Thomas Denman (1733–1815) and William Smellie (1697–1763) were physicians focused on midwifery and obstetrics whose publications defined medical literature in the generation preceding Spratt's. *The Medical Examiner: A Monthly Record of Medical Science* (Lindsay & Blakiston, 1848), 359; Lyle Massey, "Pregnancy and Pathology: Picturing Childbirth in Eighteenth-Century Obstetric Atlases," *Art Bulletin* 87, no. 1 (2005): 73–91.

12 Michael Sappol, *A Traffic of Dead Bodies: Anatomy and Embodied Social Identity in Nineteenth-Century America* (Princeton, N.J.: Princeton University Press, 2002), 192–95; Susan Wells, *Out of the Dead House: Nineteenth-Century Women Physicians and the Writing of Medicine* (Madison: University of Wisconsin Press, 2001); Susan Shifrin, "'The Worst Are Women Doctors': Nineteenth-Century Attitudes toward the Appearance and Professionalism of Women Physicians," *Transactions and Studies of the College of Physicians of Philadelphia* 5, no. 16 (1994): 47–65.

13 Sappol, *A Traffic of Dead Bodies*, 298.

14 Several period reviews echoed this idea, likely because the author himself positioned his publication for this purpose in the book's introduction. As the Philadelphia reviewers in the *American Journal of Medical Sciences* put it, the plates were useful for "recall[ing] to the mind of the young practitioner, subjects of which his recollections are liable to become obscured or confused, unless

occasionally revived by plates, models, preparations, or some appropriate apparatus." "Bibliographical Notices."

15 "Our Library Table," *Athenaeum: Journal of English and Foreign Literature, Science, and Fine Arts* 271 (1833): 794–95.

16 J. V. C. Smith, ed., *Boston Medical and Surgical Journal* 37 (December 1847).

17 "Bibliographical Notices."

18 A selection of Maygrier's publications include J. P. Maygrier, *Manuel de L'anatomiste, Ou Traité Méthodique et Raisonné Sur La Manière de Préparer Toutes Les Parties de L'anatomie, Suivie D'une Description Complète de Ces Mêmes Parties,* 4th ed. (Paris: L. Duprat-Duvurger, 1807); Maygrier, *Nouveaux Élémens de La Science et de L'art Des Accouchemens,* 2nd ed. (Paris: De Pelafol, 1817); Johann Caspar Lavater et al., *L'art de connatre les hommes par la physionomie,* New ed. (Paris: De Pelafol, 1820); Maygrier, *Nouvelles Démonstrations D'accouchemens: Avec Des Planches En Taille-Donee, Accompagnées D'un Texte Raisonné, Propre À En Faciliter L'explication* (Paris: Béchet, 1822).

19 For more on how *Nouvelles Demonstrations* (or *Nuevo metodo para operar en los partos*) fit into nineteenth-century Mexican print culture, see Nora E. Jaffary, *Reproduction and Its Discontents in Mexico: Childbirth and Contraception from 1750 to 1905* (Chapel Hill: University of North Carolina Press, 2016), 63–70.

20 Even in 1883, medical reviewers recognized when plates from Maygrier's *Novelles Demonstrations* were inserted into later publications, as did one reviewer of Dr. A. Martin's *Atlas of Gynecology and Obstetrics* published in the *Buffalo Medical Journal* in 1883.

21 J. P. Maygrier, *Midwifery Illustrated,* trans. Augustus Sidney Doane (New York: J. K. Moore, 1833), 42.

22 Maygrier, 43.

23 Maygrier, 44.

24 Tom Gunning, "The Play between Still and Moving Images: Nineteenth-Century 'Philosophical Toys' and Their Discourse," in *Between Stillness and Motion: Film, Photography, Algorithms,* ed. Eivind Røssaak, 27–44 (Amsterdam: Amsterdam University Press, 2011).

25 Maygrier, *Midwifery Illustrated,* 45.

26 Lisa Forman Cody, "The Politics of Reproduction: From Midwives' Alternative Public Sphere to the Public Spectacle of Man-Midwifery," *Eighteenth-Century Studies* 32, no. 4 (1999): 477–95.

27 Marcia Nichols, "The Man-Midwife's Tale: Re-reading Male-Authored Midwifery Guides in Britain and America, 1750–1820," PhD diss., University of South Carolina, 2010, 4; Sappol, *A Traffic of Dead Bodies*; Adrian Wilson, *The Making of Man-Midwifery: Childbirth in England, 1660–1770* (Cambridge, Mass.: Harvard University Press, 1995).

28 Wilson, *Making of Man-Midwifery,* 91.

29 Nichols, "Man-Midwife's Tale," 4; Sappol, *A Traffic of Dead Bodies,* 88–95.

30 Nichols, "Man-Midwife's Tale," 4.

31 F. H. Ramsbotham, "Dr. F.H. Ramsbotham on the Practice of Midwifery," *London Medical Gazette; or, Journal of Practical Medicine* 13 (April 1833): 886.

32 See Johns on the physiology of reading. Adrian Johns, *The Nature of the Book: Print and Knowledge in the Making* (Chicago: University of Chicago Press, 1998), 380.

33 William H. Sherman, "Toward a History of the Manicule," in *Used Books: Marking Readers in Renaissance England,* Material Texts (Philadelphia: University of Pennsylvania Press, 2009), 29.

34 E.g., Elisabeth Bronfen, *Over Her Dead Body: Death, Femininity, and the Aesthetic* (New York: Routledge, 1992); Luce Irigaray and Edith Oberle, "Is the Subject of Science Sexed?," *Cultural Critique,* no. 1 (1985): 73–88; L. J. Jordanova, *Sexual Visions: Images of Gender in Science and Medicine between the Eighteenth and Twentieth Centuries,* Science and Literature (Madison: University of Wisconsin Press, 1989). It is important to note that these works have not been without critique within their own fields. I cite them here as germinal examples of how anatomical models have entered into theoretical discourses, a legacy upon which my work depends but not an area to which I primarily intend to contribute.

35 The echoes in reception of nude marble sculpture and scientific models of the female body are a subject of my ongoing research on this topic, initially explored in my dissertation. Juliet S. Sperling, "Animating Flatness: Moving Images in American Art, 1780–1895," PhD diss., University of Pennsylvania, 2018.

10

Models and Manufactures: The Shoe as Commodity

LISA GITELMAN

When Philadelphia held a giant parade to celebrate the U.S. Constitution in 1788, three hundred shoemakers marched together in front of other tradesmen. Six abreast in matching leather aprons, they followed an elaborate horse-drawn parade float, a model "cordwainer's shop, in which were six men actually at work: the shop hung round with shoes." That may seem like a lot of shoemakers at once, but by 1850, the largest firms in Massachusetts would each employ more than three hundred men and women. In 1860, the state of Massachusetts alone had some sixty-three thousand workers employed making shoes, and shoemaking occupied more people in the United States than any industry except agriculture. Even though it lagged behind textiles as the site of industrial revolution, shoemaking would eventually come to be understood as a cardinal instance of making: Philadelphia's parade had its model shop; the nation would have its model industry. One early observer of American business history writes, "The evolution of industrial organization finds here an unusually complete illustration." The great labor historian John R. Commons had already said the same thing, if on significantly different terms. For Commons, shoemaking was exemplary because it illustrated organized labor from its colonial roots to the industrial trade unions of the later nineteenth century.[1]

While shoemaking thus became a model industry for students of American history, it was also an industry intricately structured by coeval vectors of modeling and mechanization. Starting with the wooden models—the "lasts"—on which shoes were made, this chapter examines nineteenth-century shoe production as an arena in which the idea of the model can be rendered legible in detail. Shoemaking not only depended upon models, and models of models, but the factory-made shoe itself became an architecture on which labor, industry, person, and locality were reimagined. Before industrialization, shoes were local products, and shoemakers knew the feet their shoes were made for. After industrialization, shoes were non-local products mass-produced for an impersonal marketplace. The shift in scale was one that depended at multiple junctures upon the diverse application of modeling as a way of making 3D sense.

To anyone who has encountered them, shoe lasts are self-evidently models. They look like feet. Shoe lasts are three-dimensional forms on which shoemakers construct shoes. Once made of wood, today they are often made of plastic. The *Oxford English Dictionary* traces the last all the way back to Old English, when the term (various spellings) referred to a footprint—that meaning now obsolete—and to a "model of the foot used in shoemaking." The self-evidence of lasts as models resides both in their visual resemblance to feet and in their abstracted forms, since abstraction is one route by which modeling addresses its specific functions. Shoemakers don't need lasts to look or work exactly like feet, after all, only to present stable volumetric forms that roughly correspond to the shapes and sizes of feet so that shoe parts can be patterned and assembled on them. Call this Modeling 101 or simple one-to-one mapping at scale in 3D: big lasts mean big shoes for big feet, narrow lasts mean narrow shoes for narrow feet, and so on. But shoe lasts turn out to be a little bit more complicated than that. In practice, shoemakers recognized that the last had to be a "faithful copy of the foot" in some dimensions, but in others the shape of the last had to account for the shoe it was going to be used to make: the way the shoe would position the foot, but also the way the shoe would be put together and the way it would stretch and wear.[2] Different design elements—the height of the eventual shoe's heel, for instance; the stiffness of materials; the method of assembly—have a bearing on the shape of the last. So lasts are models, yes, but because they are tools used in shoemaking, what they model are simultaneously feet and shoes. Chicken and egg, they model the foot for the shoe and the shoe for the foot (Figure 10.1).

Even apart from this double identity, shoe lasts have a lot to tell us about the connection of modeling and making in the modern period because lasts themselves were a remarkably early article of industrial manufacture. Starting in the 1810s, last making "engines," as British savant Charles Babbage called them, were designed to turn out lasts.[3] These engines were ingenious power lathes that could be used to carve the highly irregular forms that had previously been shaped laboriously by hand. American inventor Thomas Blanchard devised the most successful version, originally for making gunstocks at America's federal armories. Blanchard sold the government on his idea by bringing "a moddle"—as he spelled it—of the lathe he proposed to build to the Springfield Armory in Massachusetts, where he was ultimately commissioned to construct a full-sized version. The patent Blanchard received in 1819 did not mention lastmaking, but the reissue he got in 1820 did, after he had made a study of competing devices.

Today, if you were going to use a lathe to turn gunstocks or shoe lasts—or

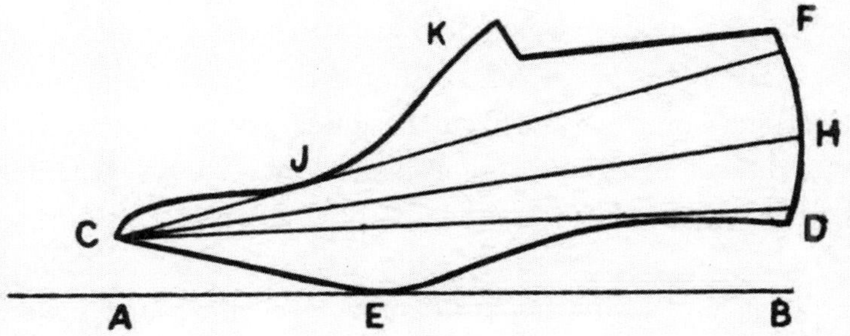

FIG. 11.

FIGURE 10.1. "Describing the different dimensions of a shoe last," in Frank Plucknett, *Introduction to the Theory and Practice of Boot and Shoe Manufacture* (London: Longmans, Green, 1916), 17, figure 11.

hat blocks or axe handles—a computer might direct the cutting tool across the wooden workpiece in relation to a digital model. Blanchard's invention guided the cutting tool in relation to a physical model. The basic principle was a sort of 3D pantography: an existing or original last was loaded into the machine and acted as a cam, a rotating form that worked as a mechanical linkage, its irregular shape guiding the cutting tool of the lathe as it passed over the surface of a spinning workpiece to produce a new last. The kinematics of Blanchard's machine—the way parts moved—was original. Improved versions of Blanchard's lathe could produce multiple lasts from a single model, could eventually produce different sized lasts from the same model, and could produce matching right and left lasts for making right and left shoes. Thus, at the same time that they modeled feet and shoes, after Blanchard and his competitors had refined the mechanism, lasts themselves were made based on models.[4]

These details about Blanchard are instructive in part because they affirm the role of physical models in the conception and demonstration of new technology. Blanchard used a model of his lathe as a visible, tangible part of his rhetorical appeal, convincing the armory superintendent at Springfield to adopt his methods. Like an architect's scale model, the model lathe showcased his design. Whatever he actually said in displaying his model, his words would have appealed mostly to the superintendent's reason *(logos),* while the model he presented appealed as well to the superintendent's trained eye and his know-how *(technē).* Blanchard would also have needed a model and a written specification to receive his patent from the federal government, and it seems he also displayed a working model to manufacturers in Boston.[5] (No model survives that I know of.) In all of these venues, his model pointed prospectively toward full-size lathes, yet in different venues, modeling had different functions. The same model meant different things to different audiences as the ways in which models have meaning changed in accordance to different contexts. Displayed to the superintendent or to businessmen, Blanchard's model suggested the feasibility of his design, arguing that the lathe would actually work and that its working would be valuable in the anticipated contexts. Received by the Patent Office, however, Blanchard's model instead suggested the patentable *idea* of Blanchard's lathe, arguing its novelty and non-obviousness. In the former instance, the meaning of the model was predictive: a "proof of concept," we might say today. In the latter instance, the meaning of the model was more analytic, establishing the idea to which Blanchard had claim. It was this idea—Blanchard's intellectual property—that would be embodied in subsequent lathes and tested in courtrooms over the ensuing decades.

Blanchard's model did its several jobs, earning him contracts, garnering a patent, and attracting licensees. Yet the patented idea that was embodied in lathes and licensed to capitalists proved remarkably changeable over time. This is where patentable ideas and architectural designs diverge. Although his patent model remained the same, Blanchard's idea continued to morph, and (lucky for him) it expanded over time in the eyes of the law, as Blanchard had his patent reissued (revising its claims), as Congress granted him two extensions, and as he vigorously pursued litigation to protect his invention against infringement. In the antebellum American patent regime, models were effectively a "medium in which the embodiment [of an idea] was anatomized and interpreted," necessary to the patenting process and then available to jurisprudence in the adjudication of infringement.[6] (See chapter 6, by Reed Gochberg, in this volume.) Models were often used to help decide when two technologies were the same (infringing) or different (non-infringing). Reinterpreted in the courtroom via a model, the particulars of Blanchard's invention were subject to change, and—in a stunning reversal—his idea ultimately expanded to include a competing lathe technology from which he had actually distinguished his version thirty years before. Whereas Blanchard had once articulated a salient difference between his lastmaking machine and one in operation in Waterbury, Connecticut, later adjudication determined that the two lathes embodied the same idea, and (happily for him) it was Blanchard's.[7]

Blanchard cuts an important figure in the history of industrial production mostly because the gunstocking version of his lathe was eventually one of sixteen special-purpose machines he designed to be used sequentially in the manufacture of small arms at the Springfield Armory. The Springfield production line became a showpiece for visiting dignitaries, illustrating the division of labor across machine tools and expressing the rational ideal—not yet fully realized—of complex assemblies made with interchangeable parts. This was "the American system of manufactures," and today it is seen as a prominent landmark along the pathway to later mass production methods, that is, to Fordism.[8] If locks, stocks, and barrels could each be made uniform, then they would be interchangeable across rifles, and rifles would be interchangeable across soldiers. Soldiers by these lights are assumedly interchangeable to begin with, and armies and navies were for the same reasons early markets for ready-made clothing and footwear.

Adapted to lastmaking, Blanchard's lathe was not part of a production line, so it did not whisper of the future with as much celerity as the gunstocking version. But lastmaking engines came to Charles Babbage's attention as one illustration of a larger, general principle that he saw pervading modern manufacturing and that underwrites commodity culture to this day. Babbage called it the principle "of COPYING, taken in its most extensive sense."[9] He was referring to industrial processes like stamping, printing, punching, and casting, in which products are shaped by dies, matrices, molds, or similar special-purpose machine tools. Products made in these ways are effectively copies of the tools that form them (usually inverse copies), and once you have the tools, there can be lots and lots of copies: hence the copiousness of mass production and the industrial commodity form. (The word *copious* is from the same root as *copy.*)

Not all manufacturing-as-copying involved actual models, but some did, and in these instances, *model* meant *original.* Wedgewood porcelains were an early example, shaped and fired in molds that had been formed around wax, clay, or plaster originals, called models.[10] The lastmaking engine was a new demonstration of the same principle. It copied an original last and could turn out as many identical copies of that model as desired. "Nothing is more remarkable, and yet less expected," Babbage enthused, "than the perfect identity of things manufactured by the same tool."[11] Machine-turned lasts achieved uniformity that their hand-carved predecessors could not. There was no assembly of interchangeable parts involved, but uniform—interchangeable—lasts were to have a profound effect on shoemaking and thus on shoemakers, and thus also on feet.

Before we return to shoemaking, however, it would be well to reflect how far the shoe last has taken us beyond Modeling 101. Our supposed curriculum began with the model as a 3D mapping of some physical form, an abstracted version, possibly altered in scale. Modeling of this sort works as a nonverbal knowledge practice, we might generalize, a way of knowing that allows someone to make 3D sense of a specific referent in a specific setting. But neither lasts themselves nor Blanchard's "moddle" functions simply in this way. Because they are tools used in making shoes, lasts must model both shoe and foot, making 3D sense of each mutually in relation to the other during the process of making shoes. And because it is used agonistically—to argue over and persuade—Blanchard's model had to illustrate the patentable idea that his lathes embodied, even when lathes weren't built yet, or when they got adapted for new ends, or when potential infringements arose. The referent of the model in each case is more difficult to pin down than we might have anticipated, and the contexts of manufacturing add new wrinkles, since the way of knowing I have just called "making 3D sense" is clearly a process distributed over space and through time, one involving materials, tools, and labor. In practice, a shoemaker makes 3D sense of leather parts tacked to a last/model as one step in the skilled labor of making a shoe. The operator of a lathe, meanwhile, ostensibly makes 3D sense of its spinning workpiece via the cutting tool and in reference to the cam/model, producing a new last. In the latter instance, a good bit of sense making has been delegated in advance to the lathe and thanks to Blanchard. The mechanized production of lasts thus exhibits what Karl Marx terms an objective principle: the craftworker has become a mere operator, "separated from the activity of conceiving and fabricating the product."[12]

Admitting this focus on conception plus fabrication, Modeling 102 might begin comparatively, by thinking about models as variations on the 2D inscriptions that sociologist of science Bruno Latour explores in his classic paper "Visualization and Cognition: Drawing Things Together."[13] Inscriptions for Latour are "immutable mobiles," like ink blots recorded by a laboratory instrument during an experiment, tables and diagrams that interpret those blots, and the published article that follows. Inscriptions accumulate in cascades like this, progressively more removed from the phenomena they mobilize, and they help to give scientific findings their potential power. Inscriptions are themselves mobilized variously as resources—as allies, Latour puts it—by scientists trying to make a convincing point. And it's not just scientists who use inscriptions, of course; they are ubiquitous and vernacular instruments in the scenography of domination broadly understood. It's not that paperwork *is* power; it's that power entails adept and diverse mobilizations of paper or its surrogates.

The models I am concerned with here aren't inscriptions, of course. They aren't flat, and they definitely cannot be incorporated into a written text. They aren't verbal, and they also aren't graphical. But they are immutable mobiles, and they share certain powerful advantages with inscriptions: they are stable, portable, and comparable, and they can be rescaled and combined—all virtues that Latour itemizes. They may also accumulate in cascades. Babbage was interested in copying, not models, but he did notice the way that copying processes sometimes reinforced each other in a linked chain. Taking the example of printing as a form of manufacture, Babbage explains that a page is copied (imprinted) from a stereotype plate, itself copied (cast) from a plaster mold, which had been copied from a typeset form: a copy of a copy of a copy. He uses his own book as an example, but he might also have used his own shoe. Shoe leather is cut—that is, copied—from patterns. In Babbage's day, wet paper was molded to fit the last, and when it was dry, the paper was cut to form 2D pattern pieces. These patterns could then be copied in leather or metal to make more durable exemplars, and then shoe leather itself was copied—that is, cut—according to these patterns. Shoe parts were effectively copies of copies. Hand cutting eventually gave way to die cutting (using a prefabricated, custom-shaped cutting tool), adding yet another link (making the tool) in the chain of copying and another opportunity for rationalizing production.

Reusing the same patterns with appropriate lasts meant making "the same" shoe in multiples. Uniformity was remarkable to Babbage and his contemporaries. It was also clearly transformative, and machine-turned lasts illustrate this point. "Uniform lasts were in an important sense not only tools," Carolyn Cooper notes, but effectively became "the gauges for achieving uniformity of production."[14] Machine-turned lasts were made in batches, and each batch contained individual lasts of the same shape and in a specific size. With standardized lasts, shoes could finally come in standard sizes; they were made "to fit no particular foot but [rather] a conception of the foot."[15] Shoes and shoemakers both began to generalize their subjects as increasing "standardization transformed the conception of the product and its production."[16] It was as if the abstracted form of the last as a model of a foot (and shoe) had finally spilled over and abstracted feet (and shoes) in general, enabling the modern shoe industry and the mass-produced shoe. Lasts had always been models for shoes, but after Blanchard's lathe made lasts into models for other identical lasts, shoes were made in multiples of a uniform style and size. Shoes became widgets, at least to the companies that produced them and to later students of business and economic history.[17] Like inscriptions operating in the 3D space of a manufactory, gauges, templates, and models can be understood as instruments in a now rescaled scenography of domination, here in the organization of industrial production and operations on the shop floor.

In shoemaking, as in other industries, standardization enabled—and was supported by—structural features that can be described with further recourse to models and modeling. Most straightforwardly, standardized production meant that selling—wholesaling and later retailing—could happen in relation to samples, that is, to models. Jobbing houses could display and drummers could carry model shoes for which they were prepared to fill orders in bulk, while retail stores can and do display sample shoes, confident that "the same" shoes can be kept in stock. In sales and promotion, as in fashion, the model shoe works as a representative of its type, standing in for all shoes of the same design or make. A wholesale buyer looks at samples and guesses which will attract consumers. A shopper picks up a sample shoe at the store and wonders what it would look like in her size and on her feet. Both encounters depend upon a shopper's tacit understanding of her own feet as more or less generic bearers of the represented shoes, rather than as dimensionally idiosyncratic body parts that must be accommodated by a bespoke product. This, then, is another spillover or cascade effect: the deskilled labor that turns the worker into something like a machine has its afterimage in the generic forms of ready-made clothing and footwear that turn the consumer into a denizen of commodity culture.[18]

The model as sample or representative is notably different from the earlier conception of models as originals, such as the last/cam in Blanchard's lathe on which further lasts were based or the wax mock-up used to create a mold for making porcelains. The model-as-representative more closely resembles Blanchard's model lathe, helping to make an appeal aimed at prospective sales, but unlike the "moddle," it belongs unequivocally to the copious—the profligate—domain of mass production and commodity culture. Here the model-as-representative aligns with the work of related 2D inscriptions: the shopper is hailed by illustrated representations of representative shoes in magazines, catalogs, and online and may imagine what those doubly represented shoes would be like in 3D.

Wholesaling was only one element in the radical restructuring of shoemaking during the nineteenth century; the division of labor was another. Old-fashioned cordwainers (and their descendant makers of bespoke footwear) made entire shoes for the specific feet they had measured, but with increasing standardization, different workers, even in different locales, could now handle making the different parts of a shoe—the upper and bottom—and its final assembly for sale to far-flung buyers and for unknown, unknowable feet in an impersonal marketplace. Massive shipments of cheap brogans (ankle-high work shoes) left New England, Philadelphia, and New York to supply southern plantations—shoes for slaves—and western frontiers, before ready-made shoes were adopted broadly by elites. Even if mechanization lagged, the division of labor objectified: "As late as 1851 all of the labor in the manufacture of shoes was hand labor," yet the individual worker "is made the automatic motor of a fractional operation."[19]

Eventually, some jobs were done in central shops, while others were "put out" to small domestic workshops and their products returned to the central shop for inspection and finishing. Next came full-fledged factories, and then "the factory system was followed by [its] outward sign, i.e., power machinery."[20] The actual "making" of the shoe—mid-century trade jargon for attaching the upper and bottom—proved something of a bottleneck into the 1860s, when the McKay sewing machine was introduced and then tested in the production of "strong army shoes" for Union soldiers.[21] The effect, John R. Commons writes, was to usurp "not only the highest skill of the workman but also his superior physique."[22] The McKay machine required eye–hand coordination, not physical strength. "Welting" and

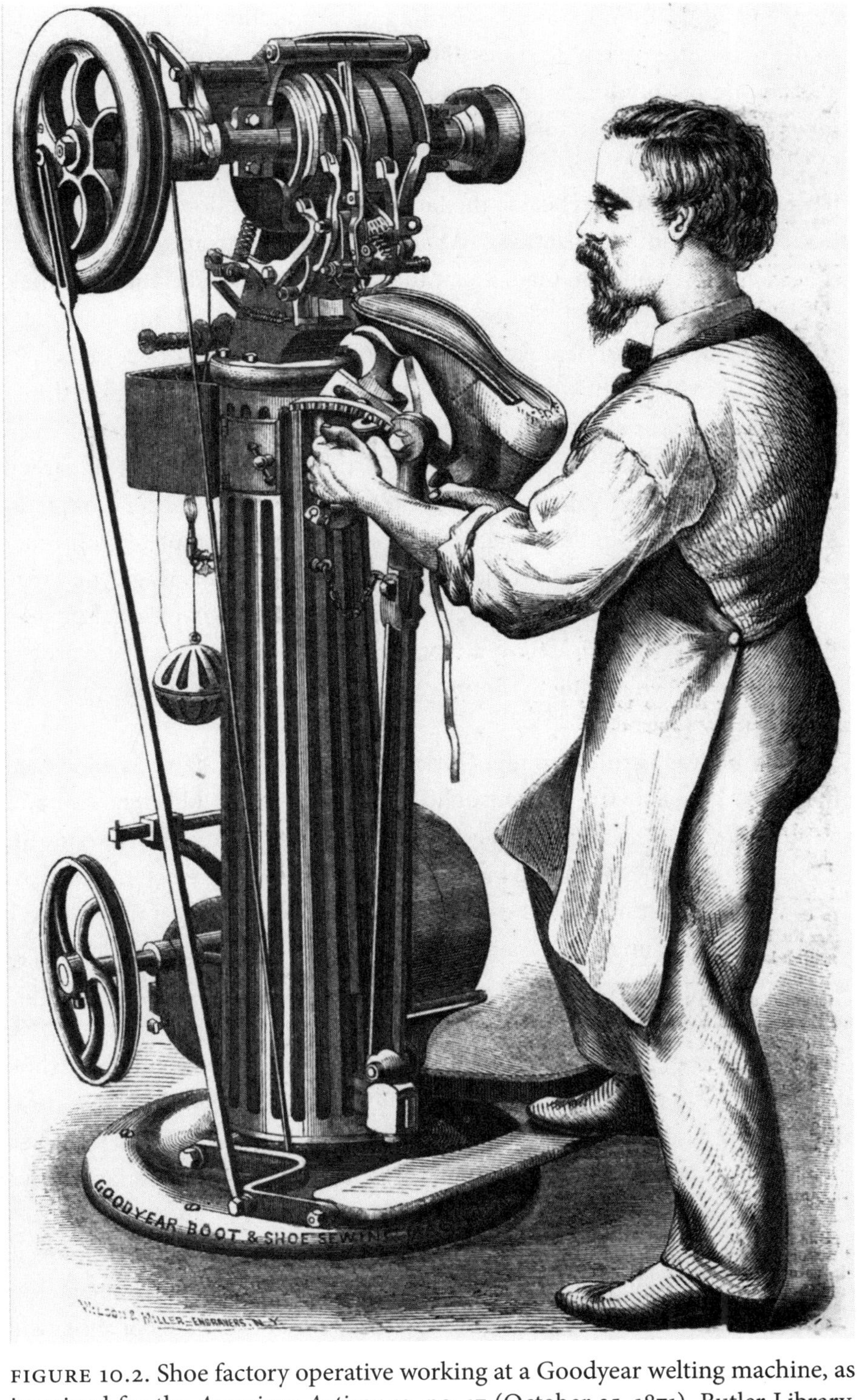

FIGURE 10.2. Shoe factory operative working at a Goodyear welting machine, as imagined for the *American Artisan* 13, no. 17 (October 25, 1871). Butler Library, Columbia University in the City of New York.

"lasting" machines soon followed, speeding up production and further reducing the skills needed to assemble shoes. The seated artisan hammering sole leather on a "lapstone" or holding a half-made shoe on his knee with his "stirrup" had given way to the trained factory operative standing at a special-purpose machine (Figure 10.2).[23]

These cursory details rehearse the familiar transformation of labor by capital in the process of industrialization.[24] As shoemakers rose from their benches and entered factories to operate machinery, the four steps in traditional shoemaking became thirty-three distinct jobs by 1880 and would be at least one hundred jobs by 1920.[25] The little old shoemaker became a relic, a philosopher's romance.[26] But what of models? Is it right—or is it helpful—to apply the word *model* in this context? Going from four steps to thirty-three or one hundred certainly reconceived (remodeled?) shoemaking and thus shoes as articles of manufacture. Clearly new production methods required new (model) workers. The model worker is an ideal type it seems, an interchangeable factory operative complete with reduced physique, upright posture, automatic motions, limited skill, low wages, and regular hours. This is *model* in quite a different sense than model-as-original or model-as-representative, since modeling a worker through his or her work processes involves a kind of structuring figuration, the model as normative outcome proper to a disciplinary regime.

Though he was hardly a model worker, America's most famous shoe factory operative was probably the anarchist Nicola Sacco. Sacco and his fellow anarchist Bartolomeo Vanzetti became a cause célèbre after they were convicted of murdering a shoe factory paymaster and guard and sentenced to the electric chair. (After lengthy appeals, both men were executed in 1927.) Before any of this, Sacco had been an edger, using an edging machine in a factory. Idle and frustrated in prison, he wrote to his English tutor, "I am joy whin [*sic*] I am work."[27] Sacco was confined without employment, although it is worth noting here that before American prisoners "made license plates," as they are said to do today, many of them made shoes.[28] Another notable in the history of shoemaking marks an earlier phase in industrial evolution with additional contradiction: the entrepreneurial Henry Wilson—future vice president under Ulysses S. Grant—made a fortune manufacturing and wholesaling shoes for slaves before the Civil War, despite his antislavery politics as a senator from Massachusetts.[29] Prison labor and southern supply remind us how the penitentiary and plantation have rightly been seen as cousins to the factory.[30]

We are left to wonder briefly (Modeling 103?) about the utility of shoes themselves as models. The symbolic meanings of shoes are of course legion,[31] but what could a shoe be said to model, even if this use of the term walks us toward the realm of metaphor and fancy? Is it right—or is it helpful—to say that shoes model feet and ground in relation to each other, making 3D sense of both together? Nineteenth-century slang implied as much, since shoes and boots were sometimes punningly called "understandings."[32] Or perhaps it's better to say that shoes model distance as a social and a geographical fact.[33] Nineteenth-century sources like Henry Mayhew's *London Labour and London Poor* (1852–61) confirm the extent to which social distance was casually descried in distinctions between the well shod—the polished—and the slipshod, as well as between the shoeless and the merely down at heels. And while shoes certainly signal the fortunes of the persons who wear them, they also signal the geographies of production.

Factory-made shoes displaced handcrafts in an uneven, century-long process that spanned the nineteenth century as it spanned the globe. Richard Henry Dana's immensely popular *Two Years before the Mast* (1840) offered its early readers a glimpse at the transition already well under way. Dana had sailed to California on a merchant ship from Boston. On the outbound voyage, their cargo included "boots and shoes from Lynn, calicoes and cotton from Lowell." The return cargo was hides, which Dana and his crewmates collected up and down the West Coast and painstakingly processed for the voyage home. Twice in the course of his memoir (lots of things happen twice in *Two Years before the Mast*), Dana imagines that the hides are taken to Massachusetts, tanned, and made into shoes to wear on feet in California to collect hides taken to Massachusetts, tanned, and made into shoes to wear on feet in California, and so on.[34] Dana's vision reads like an update of the Grimms' fairy tale tailored to the mill towns of New England. In "The Elves and the Shoemaker," elves help a poor but charitable old shoemaker who is down to his last bits of leather. In *Two Years before the Mast,* production and supply are recursively related and enduring; factory hands are busy far away.

The Grimms' shoemaker has elves as allies. I have argued here that shoemakers' lasts are no less allies: they are epistemic resources in the work of making 3D sense of shoe leather. Elves are magic. Shoe lasts are models. Seen within the contexts of industrialization, shoes and lasts point to modeling as a diverse array of processes integral to making and manufacturing. In the very least, models can be originals (like the cam/last in Blanchard's lathe); models can be representatives (like a drummer's sample or a retailer's display); and models can be ideal types, like workers acquiescent to new tools and factory discipline. Even apart from this diversity, models and modeling make challenging subjects of inquiry. Whether the lasts (modeling foot *and* shoe), Blanchard's moddle (modeling his lathe *and* the changeable idea that it embodied), or other instances, it seems it can be difficult to determine referents—what a model is *of*—and so what the word *model* actually means.

NOTES

1 Blanche Evans Hazard, *The Organization of the Boot and Shoe Industry in Massachusetts before 1875* (Cambridge, Mass.: Harvard University Press, 1921), v; John R. Commons, "American Shoemakers, 1648–1895, a Sketch of Industrial Evolution," *Quarterly Journal of Economics* 24 (November 1909): 39–74.

2 Frank Plucknett, *Introduction to the Theory and Practice of Boot and Shoe Manufacture* (London: Longmans, Green, 1916), 18.

3 Charles Babbage, *On the Economy of Machinery and Manufactures,* 4th ed. (1835; repr., New York: Augustus M. Kelley, 1971), 102.

4 All details from Carolyn C. Cooper, *Shaping Invention: Thomas Blanchard's Machinery and Patent Management in Nineteenth-Century America* (New York: Columbia University Press, 1991); "moddle" appears on page 87.

5 Cooper, 172.

6 Alain Pottage and Brad Sherman, *Figures of Invention: A History of Modern Patent Law* (Oxford: Oxford University Press, 2010), 16.

7 Cooper, *Shaping Invention,* 52, 54.

8 David A. Hounshell, *From the American System to Mass Production, 1800–1932* (Baltimore: Johns Hopkins University Press, 1984), 21, 35, 331–36.

9 Babbage, *On the Economy of Machinery and Manufactures,* 69.

10 See Malcolm Baker, "Representing Invention, Viewing Models," in *Models: The Third Dimension of Science,* ed. Soraya de Chadarevian and Nick Hopwood (Stanford, Calif.: Stanford University Press, 2004), 28.

11 Babbage, *On the Economy of Machinery and Manufactures,* 66.

12 Ross Thomson, *The Path to Mechanized Shoe Production in the United States* (Chapel Hill: University of North Carolina Press, 1989), 1.

13 Bruno Latour, "Visualization and Cognition: Drawing Things Together," *Knowledge and Society Studies in the Sociology of Culture Past and Present* 6 (1986): 1–40. Latour is also a point of reference in Chadarevian and Hopwood's *Models: The Third Dimension of Science.*

14 Cooper, *Shaping Invention,* 180.

15 Thomson, *Path to Mechanized Shoe Production,* 36.

16 Thomson, 36.

17 In the period between Commons's and Hazard's accounts of the industrial development of shoemaking, the Harvard Business School used shoe selling as an initial case; see Harvard University, "Object and History of the Bureau in Brief, with Some Preliminary Figures on the Retailing of Shoes," *Bulletin of the Bureau of Business Research* 1 (May 1913).

18 The volume editors were helpful in making this last point. For more on ready-made clothing, see Richard Sennett, *The Fall of Public Man* (New York: Vintage Books, 1978), 147ff.

19 Commons, "American Shoemakers," 72; Marx, *Capital* (1867), vol. 1, chapter 14, section 5, available at http://marxists.org/archive/marx/works/1867-c1/ch14.htm#40a. Commons is overlooking innovations of the 1840s like the hand-powered skiving machine and machines for cutting and rolling sole leather. Hazard, *Organization of the Boot and Shoe Industry,* 93. Marx is not writing specifically of shoemaking but of the division of labor in manufactures.

20 Hazard, *Organization of the Boot and Shoe Industry,* 119.

21 See Hazard, 96. Also see Thomson, *Path to Mechanized Shoe Production,* 38ff.; David N. Johnson, *Sketches of Lynn; or, The Changes of Fifty Years* (Lynn, Mass.: Thos. P. Nichols, 1880), 342. On army shoes, see *Manufactures of the United States in 1860; Compiled from the Original Returns of the Eighth Census, under the Direction of the Secretary of the Interior* (Washington, D.C.: Government Printing Office, 1865), lxxii.

22 Commons, "American Shoemakers," 73.

23 On the desire to have shoemakers stand, not sit, see *Manufactures of the United States in 1860,* lxx: "The London Society of Arts, in 1802, and again in 1812, awarded premiums for machines to enable shoemakers to work in a standing position, thereby relieving the pressure upon the breast and the constraint of position, which were so detrimental to health."

24 For these oversimplifications, I rely on Commons, "American Shoemakers"; Hazard, *Organization of the Boot and Shoe Industry*; and Thomson, *Path to Mechanized Shoe Production,* all of whom offer many necessary details.

25 Hazard, *Organization of the Boot and Shoe Industry,* 3; her thirty-three jobs comes from Johnson, *Sketches of Lynn,* 344.

26 As Andrew Parker puts it, "the shoemaker is a figure against whom philosophy constitutes itself"; see "Editor's Introduction: Mimesis and the Division of Labor," in Jacques Rancière, *The Philosopher and His Poor* (Durham, N.C.: Duke University Press, 2003), xi.

27 Nicola Sacco and Bartolomea Vanzetti, *The Letters of Sacco and Vanzetti,* ed. Marion Denman Frankfurter and Gardner Jackson (1928; repr., New York: Penguin Books, 2007), 6. The letter is from 1922.

28 In *Life of P. T. Barnum* (1855; repr., Urbana: University of Illinois Press, 2000), Barnum describes a youthful visit to New York in 1822: "My friend also took me out of town to see the State Prison, paid my way in, and witnessed my astonishment at seeing so many wicked convicts dressed in the striped suit, and especially to see about two hundred shoe-makers turn their faces to the door when we entered, with as much precision as if they had been automatons all moved by a single wire" (26).

29 Hazard, *Organization of the Boot and Shoe Industry,* 69–70.

30 See Dario Melossi and Massimo Pavarini, *The Prison and the Factory: Origins of the Penitentiary System,* trans. Glynis Cousin (Totowa, N.J.: Barnes and Noble,

1981). On plantations and factories, see Sven Beckert and Seth Rockman, eds., *Slavery's Capitalism: A New History of American Economic Development* (Philadelphia: University of Pennsylvania Press, 2016).

31 As Patricia Crain notes, shoes may even be said to describe that sweet spot where the commodity fetish and the sexual fetish come together as one. Crain, *Reading Children: Literacy, Property, and the Dilemmas of Childhood in Nineteenth-Century America* (Philadelphia: University of Pennsylvania Press, 2016), 25.

32 *The Slang Dictionary: Etymological, Historical, and Anecdotal,* Rev. ed. (London: Chatto and Windus, 1874). *Understandings* means feet or boots: "Men who wear exceptionally large or thick boots are said to possess good understandings" (333). More serious is this Buddhist teaching, from Shantideva, *A Guide to the Bodhisativa's Way of Life,* which Patricia Crain pointed out to me: "Where would I possibly find enough leather / With which to cover the surface of the earth? / But (wearing) leather just on the soles of my shoes / Is equivalent to covering the earth with it. // Likewise it is not possible for me / To restrain the external course of things; / But should I restrain this mind of mine / What would be the need to restrain all else?"

33 Ann Rosalind Jones and Peter Stallybrass, *Renaissance Clothing and the Materials of Memory* (Cambridge: Cambridge University Press, 2000); clothing is "a world of social relations put upon the wearer's body" (3). Henry Mayhew's *London Labour and London Poor,* 4 vols. (1861–62), is fully searchable at Gutenberg.org.

34 Richard Henry Dana Jr., *Two Years before the Mast* (1840; repr., New York: Dodd, Mead, 1946), esp. chapters 13 and 19.

11

Modeling Interpretation

JOHANNA DRUCKER

THE CONCEPT OF A "MODEL"

Can interpretation be modeled? To answer this question, we have to define what we mean by *interpretation* and how a model is to be understood. In a vernacular sense, we use the term *model* as both noun and verb: to designate a schematic idea or construct or to indicate the act of creating a model. A model is a generalizable expression or working plan, as in a data model or a model of a strategy.[1] It might be created for a specific instance—such as a model of a particular house, city, or ship (almost all material artifacts are based on models—conceptual, industrial, traditional, or other—and then instantiated). But when a model is particular, the term refers to the conceptual work of form-giving: the model is the scheme that contains features, structures, organization, and other aspects of the object or process, but it is not the existing house or ship. A model also can be the design of a set of rules or procedures—as a model of behavior for a robot or a complex system. Each of these emphasizes the idea of modeling as primary constitutive act. This chapter takes up the problem of modeling interpretation within the context of digital humanities projects that make use of visualizations as part of their basic scholarly work. It proposes nonrepresentational methods as an alternative to those that use representational modes of display. Nonrepresentational approaches of visualization are not based on preexisting information but rather are primary acts of form-giving.

A model differs from a representation in several ways. Representations are surrogates that stand in for an already existing object, thing, or event. The term *nonrepresentational* was developed in geography, by Nigel Thrift (among others), to describe the way a space can be modeled from experience, rather than being represented by surrogates, such as maps, that assume they can adequately portray existing space or territory.[2] A nonrepresentational approach does not preclude the use of an image to depict or figure something; it simply suggests that an image cannot represent something in a relation of equivalence. The distinction between representational and nonrepresentation approaches is not made on the basis of characteristics of documents, artifacts, presentations, or expressions in graphical or textual form; rather, it depends on whether the visualization is assumed to be a surrogate for the phenomena it references.

NONREPRESENTATIONAL APPROACHES

Nonrepresentational approaches are fundamentally constructivist. A nonrepresentational presentation of a space is a primary expression of experience. By using this approach as the basis of graphical and textual expressions of data or interpretation, we can shift from positivist and empirical conventions to subjective ones. For instance, it is a truism that a map is not equivalent to a territory and that every cartographic projection system is a distortion. But a floor plan or street map drawn up using standard metrics usually provokes no such criticisms—it is taken as an accurate (or adequate) representation of the space, as if the space (like a text in a naive reading) existed independently of its being experienced. A nonrepresentational approach suggests that factors like gender, race, class, age, and other demographics are active components of modeling space and should inform our depictions. Put a lost child, a threatened woman, a preoccupied young man, on the same city block. They will not merely experience the same environment *differently,* as if it could be measured apart from their perception. Each constructs a *different space* through his or her experience. How big is a space when you are three feet tall and it takes twenty steps to cross it and your view line is constricted by trash bins or cars? Or when you can cross it confidently in four strides? Are the six inches of curb that separate you from an abuser and fast traffic the same size as six inches in the middle of the sidewalk? The materials of the humanities—films, plays, novels, poems, essays, historical documents and cultural records—are filled with works that make the case that time, space, and experience are modeled from subjective, interpretative positions. These models differ from those made with the presumption of the rational, observer-independent metrics that govern standard conventions. And yet, within the digital humanities, this interpretative activity has not found explicit expression in graphical (or other) form, nor have the models (such as data models or algorithmic ones) by which interpretation is formulated been explicitly called to attention.[3]

A nonrepresentational approach assumes that there is no unmediated relation to phenomena—be they texts, landscapes, objects, or events. The question is whether the concept can be turned to reflect upon the act of modeling itself. What, for instance, would a dashboard for modeling interpretation in a digital encounter with a text (or corpus of materials) look like if it focused on assumptions (how the space is understood in phenomenological terms) rather than mechanistic procedures (how the space can be measured)? These questions will be brought to bear here on a series of project prototypes and proposals in the digital humanities.

CHALLENGES FOR DIGITAL HUMANITIES

In general, digital humanities projects have absorbed the methods of natural and social sciences, turning away from (or at least bracketing out) the crucial theoretical issues that have been central to the humanities in traditional and current hermeneutics. Digital humanities projects and platforms have, for instance, adopted conventions of representing time, space, and quantitative data using timelines, chronologies, calendars, maps, charts, and graphs that are based on rational, observer-neutral presentations of information from the empirical sciences. The first criticism to be raised against such a presumption is fairly standard within the critical history of science—that empirical methods are steeped in ideological and historical values that are anything but neutral. Attempts to rationalize time and space, or to use methods of parameterization to extract "data" from the phenomena of the social, cultural, and natural world, can be readily subjected to a critique that exposes blind spots of race, gender, and power asymmetries across all manner of divides. Reflect on the specificity of a thermal scale, Celsius, that is bounded by freezing and boiling points of water, when entire cultures had no experience of ice until a relatively recent period. How does the Celsius model of temperature work in those cultural circumstances?

The second criticism is more generative—that the capacities of digital technology to support the graphical and formal expression of affective metrics rooted in the specificities of individuated, historically and culturally located parameters have not been explored. We have not developed platforms in which to model interpretation as such, simply to display the results of interpretative work (data structures, markup, parameterization) that frequently keep their models from view.

The attempt to engage the potential for modeling interpretation has been central to my work in digital humanities since the first Temporal Modeling experiment in the early 2000s.[4] This was driven by the central conviction that humanistic expressions of temporality—in novels, plays, film, poetry, historical study, and archival work—cannot be represented in standard timelines without doing considerable violence to the complexity of experience embedded in those works. No two minutes are the same in human experience; no standard clock time can be used to measure the pause between sentences in a dramatic scene or the acceleration of events under pressure of a deadline. The opportunity to move beyond conventions for creating temporal, spatial, and quantitative representations and instead produce a platform for creating and presenting observer-dependent expressions remains present, but largely latent, within digital humanities work. One challenge is to imagine the *primitives* of an interpretative system—but what version of interpretation would this model?

INTERPRETATION AND COMPUTATION

If we model interpretation in a nonrepresentational way, then theories of hermeneutics can inform our understanding of the way models are produced in this constitutive action. The work of exegesis is always a representation—a restatement and paraphrase—of an original. By definition, such representations are never the same as their source but often obscure that distinction. Since the term *nonrepresentational* emphasizes the impossibility of equivalence, the nonrepresentational approach inverts the sequence of ontological events. This is familiar ground in literary theory based on reader response. The act of reading constructs a text; it does not reference it as a given, static, singular, preexisting entity. Reading itself is not a monolithic practice—functional, recreational, informational, devotional, and other modes could be modeled. Questions of agency also arise with regard to the design and performative capacity of modeling an interpretative environment. What idea of a reader/interpreter is modeled? How does this embody ideological and psychological ideas about the user as a subject—or consumer? Are current screen-based expressions of information, for instance, capable of producing any conditions for subjective agency as individual and historically, as well as culturally, situated? Or are the technologies of display and interaction all premised on a consumerist model in which optimization of immediate gratification is the end goal of the design and the user is always only able to perform the scripted commands? The question gets at the heart of what a model instantiates—not only as an expression of a knowledge paradigm but a model of the imagined user–producer–text system of knowledge or information. In other words, we can show "who" we are modeling in creating an interpretation, not only "what" we are modeling.

Traditional methods of interpretation do not constitute a singular, homogenous, unitary concept or practice.[5] An interpretative act might be a search for truth, for meaning, for effect and affective force. An interpretation can be done individually or in a community, as an act of divination from signs and symbols, as a scholarly exegesis of a text, or as a response to a film among a group of friends. The term has no simple, single stability in its definition—except insofar as an interpretation stands in relation to a provocation (text, image, event, performance, object) as a constitutive act. In modern hermeneutics, at least, the notion of interpretation implies a process of *understanding* in which a human agent engages in a mediating relation to a provocation and constitutes a mental, sometimes verbal, production of it through a negotiated awareness of its properties, content, meaning-producing features, and forms.[6] Within the traditions of modern hermeneutics, particularly the strain of philosophy informed by Martin Heidegger, interpretation is a constitutive act, one that makes the object of investigation through engagement.[7] In this approach, an interpretation is never the same as the original object and is an ongoing process of negotiation between perceiving consciousness and object.

Critical work of the last half-century introduced radical changes to the hermeneutic model through poststructuralist, postcolonial, queer, feminist, and other frameworks. In these frameworks, the model of subject positionality was self-consciously brought into focus as an integral aspect of interpretative practice extending phenomenological foundations into a political arena of identity politics. More recently, the appearance of new materialist and speculative realist formulations has challenged anthropocentric biases and the basic subject–object divisions governing previous traditions. The concept of "intra-action" that emerges from work by Gilles Deleuze and, more recently, Karen Barad, among others, supersedes the "modern" formulation. Speculative realism attempts to obviate the distinction between being and knowing, working against the traditional models of cognition and knowledge grounded in the model of an apprehending subject. These arguments indicate a dramatic shift in theoretical models of knowledge as knowing—dynamic systems of codependent relations among mutually constitutive entities.

The digital humanities, in part by its concessions to the mechanistic and highly formalized, procedural logics of computational methods, has not integrated these theoretical approaches into its techniques. The model of "data production" and "visualization" in current digital humanities remains circumscribed by adherence to mechanistic approaches by which data are presumably merely "extracted" from phenomena (or digital artifacts) and represented in formal expressions and graphical forms without apparent complication. In a crucial way, the digital humanities seems to have abandoned its commitments to traditions of interpretative work in the creation of its infrastructure and methods—as if processing "texts" as "data" (calculating quantities and probabilities of identical ASCII streams) were independent of interpretative influences, merely objective views of a "text" presumed to have a fully autonomous and self-evident existence.[8] What if, instead, the digital humanities were to take on the challenge of modeling interpretation instead of giving up on the possibility of formulating computational methods on a nonempirical foundation?

The challenge is to understand how to model interpretation computationally—and expose the interpretative models at work in computational processes. How might the constitutive action of modeling be externalized, given manifest form, and communicated as an entity in itself—made tractable. Through what means and by what actions can an interpretation be modeled so that it can be apprehended as a process? How do we show the ways in which data models, for instance, produce the phenomena they appear to represent? In any reading of a text, how do we show the model from which the interpretation derives? If I read with a model of melancholy in mind, I produce a different poem from the reading than if I read for metric structure or political import. How can we make these models explicit in our arguments and presentations?

Mathematical and logical models of interpretation take the form of formulas. In computational modeling, the algorithms instantiate the models—of bias, values, priorities, and other factors. These factors are quite conspicuous in predictive modeling or probabilistic projections, where they influence outcomes within the model and the implementation of policies and actions based on these outcomes (predictive policing, for instance). Neither modeling nor interpretation is exclusive to language, despite academic familiarity with linguistic models, particularly in the humanities. All symbolic expressions—gesture, sound, movement, visual, graphical, musical, spatial forms—can be put at the service of modeling. We could go further and say that their very formalization as symbols is already the instantiation of a model—to wave a greeting is to already know what constitutes such a symbolic action, the model of which we hold in mind, even as each individual act of greeting is a unique, specific instantiation that does not "represent" a greeting but enacts it.

NONREPRESENTATIONAL APPROACHES TO VISUALIZATION

With the problems of modeling interpretation in mind, we can address several nonrepresentational approaches to visualization as the means by which to model interpretation, particularly in the digital environment.[9] The term *visualization* includes construction of arguments using graphical means, creating data through graphic manipulation and direct input, and other specific techniques for supporting interpretative approaches to scholarship through a platform and interface. The goal is to offer an alternative to the literalist and positivist representational approach that dominates most data visualizations while proposing a set of possibilities for instantiating interpretative principles in graphical form. To reiterate, while the phrase *nonrepresentational visualization* might sound like a contradiction, it merely means making use of visual and graphical methods to create a model directly, rather than using a graphical expression to restate an already existing model, form, or data set.[10] A nonrepresentational image has no preexisting referent. This experiential aspect foregrounds user-dependent attitudes toward knowledge production, rather than the user-independent modes central to empirical belief systems and their practices. This shift is an essential aspect of introducing critical hermeneutics into visualization practices, making them useful methods for modeling interpretation as a user-dependent experience, rather than suggesting that an interpretative act merely apprehends an already extant object or artifact whose autonomous existence can be understood without reference to the frame of perception within which that understanding takes place.

The fundamental tenets of an observer-dependent epistemology are at odds with those of empirical sciences, with their insistence on repeatable results.[11] The claims to universality of laws in the physical and natural world, of perception, and of explanation are counter to the recognition that knowledge is partial; situated; historically, culturally, and even individually inflected. This latter position is not a blank check for wanton disregard of consensual forms of knowledge production, any more than poststructuralism was license for an "anything goes" approach to the study of texts. All acts of interpretation are constrained, but they are also, fundamentally, processual, constitutive acts. These principles, to which I will return in the conclusion to this essay, are noted briefly here to establish the intellectual framework within which my argument is made.

Begin with the notion that nonrepresentational acts of interpretation might make use of graphical means for modeling and the primacy of the visual as a space of generative and primary work also becomes evident. Not all visual images are mimetic or representational. Not all images follow after an event, object, phenomenon, or experience. Knowledge can be constructed graphically—ordering or organizing, organizing information in a visual way is a primary mode of knowledge production.[12]

To model interpretation requires either showing acts of interpretation instantiated and/or showing what the rules and procedures are by which interpretation takes shape. The work of interpretation is not merely a relation between an observing agent and an object as if a mechanical transfer takes place. Interpretation is the third space, the constituted work of that in-between-ness—or, taken further, as mentioned above, a dynamic system of codependencies.

STANDARD VISUALIZATIONS

When visualizations are produced in a standard mode, they are considered to be graphical expressions of quantitative information—representations that stand in for the quantitative information. This approach was fully implemented by William Playfair in the late eighteenth century in a series of elegantly designed plates he put in the service of "political arithmetik"—which were statistical and numeric figures about wages, trade, commodities, and other aspects of Britain's economy

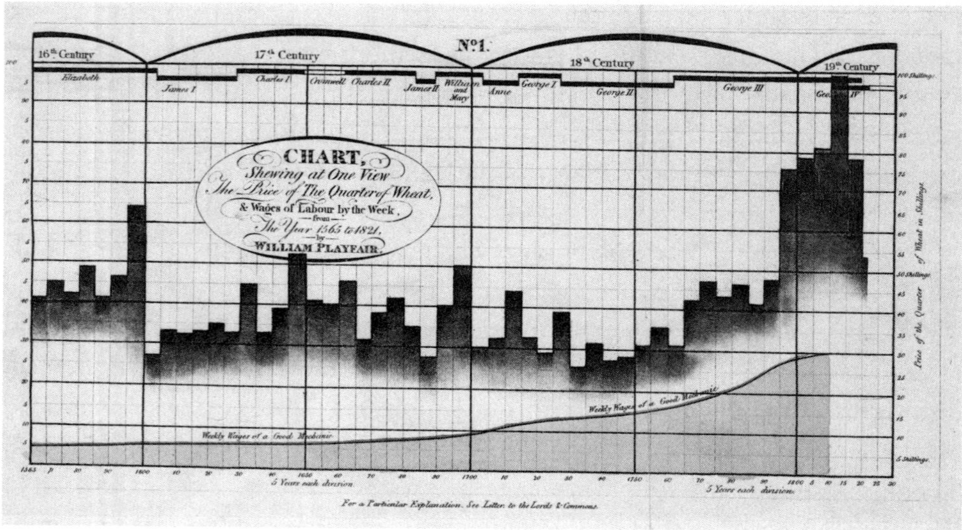

FIGURE 11.1. Playfair's chart of wheat prices and wages, in William Playfair, *A Letter on Our Agricultural Distresses* (London: William Sams, 1821). General Collection, Beinecke Rare Book and Manuscript Library, Yale University.

(Figure 11.1).[13] Playfair more or less invented the basic graphic vocabulary of bar charts and graphs. His engravings were elegant conceptually as well as visually, with multiple levels of chunking, different scales of information within a single frame, and other features. The common forms of graphical display used in contemporary visualizations nearly all have their origins within the eighteenth and early nineteenth centuries, where their rhetorical function was well recognized. Florence Nightingale, for instance, took advantage of the effect of distortions of scale. She expressed the contrast of numerical values by extending the length of a radius in a circular diagram, thus expressing comparisons as *area* (rather than length) on her charts. This dramatized her points about death rates in hospitals in contrast to those on the battlefields. Affective impact, not accuracy, was her goal, and she achieved it.[14] The techniques of production, as well as formal elements, participate in the expression of affective dimensions. The statistical charts prepared by W. E. B. Du Bois for the Paris Exhibition in 1900 were hand drawn by him and a team of his students from Atlanta University.[15] The traces of making are evident at every level of granularity—from the lettering to the pen work and coloring. This aspect of their "argument" cannot be ignored; it is visceral in its impact.

Visualizations of information or data are generally constructed through a production sequence that begins with the creation of a table of numerical values. These are then charted within a coordinate grid or chart as data points, lines, areas, or other graphical expressions. The sequence just described is linear and direct. The interpretative work of data production is displayed as a result, rather than a process. Interpretation has been modeled in every case, but the model is subsumed by the display, rather than foregrounded. To some extent, this outcome is inevitable. A static image, even an interactive interface, displays an interpretative model after its production, not in process.

AN ALTERNATIVE APPROACH: TEMPORAL MODELING

With an awareness of the limits of standard visualizations, I made an attempt to create alternatives. This work was at the heart of several projects that I worked on with my collaborators, Bethany Nowviskie and Jerome McGann, in the context of SpecLab (the Speculative Computing Lab) at the University of Virginia in the early 2000s.[16] Temporal Modeling and the Ivanhoe Game were both working prototypes that embodied theoretical principles of interpretative work.

Temporal Modeling was begun with the conviction, mentioned earlier, that standard timelines developed for the natural sciences did not work for the multiple temporalities in humanistic documents.[17] The standard features of timelines are based on assumptions that time is linear, unidirectional, homogenous, and continuous. This makes sense within an empirical framework, where the correlation of evidence and observation conforms to the assumed time-arrow and where the standard units of measure are the conventional norm. In such a framework, we do not roll time back, nor do concepts like anticipation and regret have much bearing on the conception of research. Every segment—a minute or second—is considered identical to every other second or minute, rather than being shaped by affective dimensions of experience. The format of an observer-independent system does not contain elements of subjectivity and, like classic formulations of "time" as a container, is untroubled by the distortions and complexities of discursive, narrated, or experiential time. Our goal (never quite fully achieved) was to construct a system that was not based on a container model (time as a preexisting framework into which events or incidents were put) but based on relationality. The proportions and scales of the temporal model would emerge as events and references were put into relation with each other (Figure 11.2).[18]

Unlike "time," then, temporality is not understood as a given but is constructed relationally.[19] The distinctions between time and temporality were already configured into philosophical descriptions in late antiquity, and the only major extension of temporal concepts beyond these was introduced by Albert Einstein's theories of relativity. For purposes of modeling temporality as a dimension of interpretation, however, we do not need to wrestle with the complexities of Einstein's space-time but only with those of discursive and experiential presentations.[20] In other words,

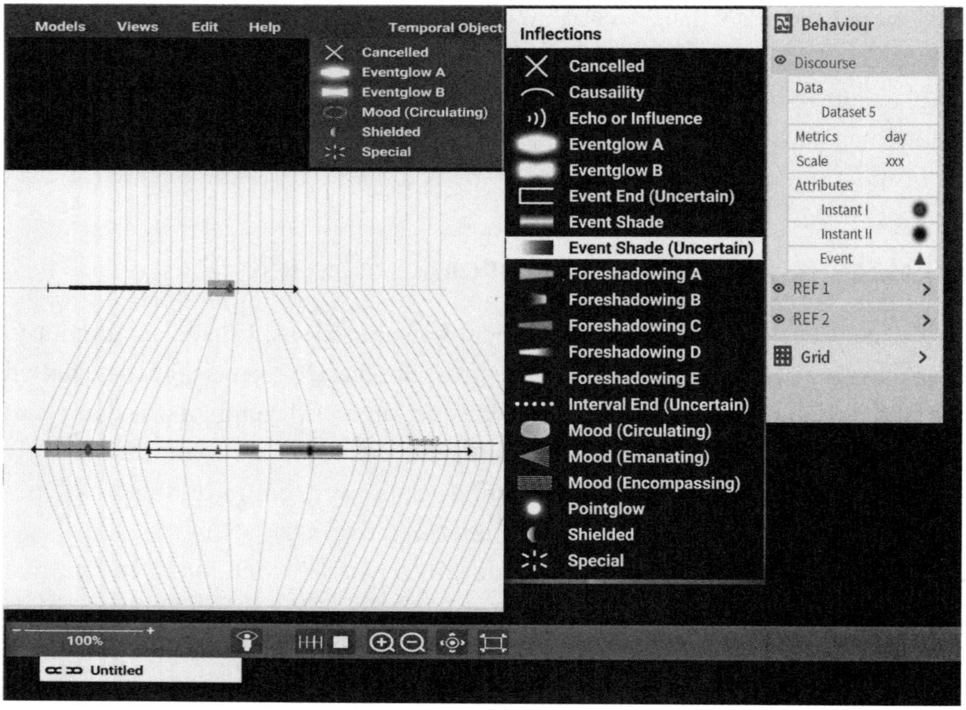

FIGURE 11.2. Temporal Modeling hybrid. Updated features created by Merle Ibach in consultation with the author.

our focus was on the way temporal relations could be modeled from texts, aesthetic works, historical documents and artifacts, and even such ongoing activities as news cycles.

In thinking about temporality in a discursive sense, the fields of humanistic expression present a number of interesting problems. One is that the temporality *of* a document and that represented *by* a document—the distinction made by narratologists between the time of telling and the time of the told—are not the same. (How long is a film by contrast to the duration of events portrayed?) This primary distinction is so familiar that any notion of "real time" within the representation of events is remarkable and exceptional, rather than the norm. No fictional account of three generations of a family's history takes sixty years to read or tell, for instance. The interpretation of temporality is presented in narrative forms that model events, intervals, and moments in unequal ways. No standard metric applies when allocating the time of telling to the time of the told. A chapter might cover a minute, or a year, and be the same length as a chapter that details events of a decade. Analyzing the temporality of a narrative is a highly interpretative act, and the time shifts within an account might involve multiple frames of temporal reference. How are anticipation and regret modeled in temporal terms? Prospective and retrospective views of an event are necessarily different. The haste with which something occurs does not match the pace of reflection on it, so different scales of temporal experience are always in play.[21]

The internal temporalities of narrative or aesthetic artifacts present one set of opportunities for thinking about how interpretation can be modeled. The task of producing a temporal model begins with reflection on its assumptions—and whether it is to be based on empirical or subjective premises. But other models of temporality arise within the humanities. Human history and cultural time-keeping methods vary across different groups and periods. The influential fourth-century historian and scholar Eusebius established a comparative chronology that charted biblical, historical, and Greek systems in relation to each other.[22] The timescales and metrics of each of these systems are distinct, different from each other even if consistent within themselves. None matches precisely the contemporary Western system of historical dating. Other methods of establishing the historical record (e.g., Chinese and Egyptian) use dynasties to periodize their dating system, or, in New World calendars, solar, religious, and astronomical systems interlock in a complex calculation of cycles and events. The seventeenth-century Irish bishop James Ussher created a chronology of the history of the world that was used well into the nineteenth century.[23] Based on the Old Testament, it corresponded to a historical understanding that dated from Creation to the present. These and other, even earlier, biblical systems of chronology were adopted by prominent figures as varied as Sir Walter Raleigh, Sir Isaac Newton, and others who calculated human history in conformance with their dating methods.[24] Any work in the humanities that depends on historical sources from this era has to model chronology on these bases—not as an incorrect understanding but as a model of chronology grounded in a specific set of assumptions. The interpretative work of the chronologies provides the foundation for the models in this instance.

Creating an interpretative framework for demonstrating these differences and modeling them in graphical form is essential in respecting the cultural otherness of the past. To do this means taking seriously the way models of human history were complete explanations of the historical past. Giving these chronologies their due, not "rectifying" them in relation to the timescale of geological time, requires a different model. The careful calculations of the many seventeenth- and eighteenth-century antiquarians whose universal histories of the world included carefully computed arithmetic on which to figure the number of years between Creation and Expulsion, for instance, should not be dismissed but presented in accord with the systems of temporal modeling on which they were based.[25] Fundamentally, this means using a timescale that expresses the model in the work, not one that shows it in relation to a "correct" timeline of which it is merely a part.

So far, we have touched on the challenges of modeling temporality as an interpretative act within narrative forms and modeling chronologies within historical systems. Other challenges for modeling temporality arise when working across a corpus of documents in any period. The floating references to "now," combined with the uneven distribution of attention to some points in a temporal sequence of events, create a shifting model of the time referenced. Consider the complex model of an event in a news cycle that gains attention over the course of several days and in very short news cycles of minutes or hours. The initial temporal interval may have been small, even trivial, but gains importance and depth through accumulated accounts. The model of the event changes shape. The narration constantly changes, and the event, as a social construction with temporal dimensions, expands elastically, even as the repleteness of information may still preserve gaps and ruptures. The event takes on new dimensions as the representation alters the model, and vice versa. Conflicting accounts of specific milestones or reference points within the event create a new model with each changed set of documents or data points. Because this is a discursive framework, the statements within the texts and documents create the temporal referent. The interpretation of the event, and its temporal shape—duration, complexity, subjective perception from multiple viewpoints—can be actively modeled (and manipulated, as we see daily). In other words, the shape of the interpretative activity—whether it is in terms of data or a graphical expression or a set of descriptive terms—can be modeled specifically and also as a demonstration of the general principles of systematically modeling such activity.

In creating the prototype for the Temporal Modeling platform, we modeled the primitives of the system as well as their behaviors and their use.[26] These primitives were *elements* (points, lines, intervals) and their *attributes* (semantic and syntactic inflections), *behaviors* (the ways the elements acted), and *actions* (what a user could do in the space). Each of these was modeled, and designed, as part of a system that was meant to support interpretation of temporal events and relations. The elements were very basic, but the idea of stretchy, expandable, elastic, or otherwise mutable timelines added a degree of flexibility within which different scales and metrics could be registered in the graphical system. In addition to allowing for nonhomogeneous scales of temporality, the system allowed for discontinuity—ruptures and breaks—as well as branching paths of possible futures and reinterpretation of past events within light of new information or perspectives. Semantic inflections allowed for characterization of any element through additional attributes—importance, vagueness, labels, and so forth—while syntactic inflections linked components in a causal or dependent relation (e.g., foreshadowing links one event and another across an interval). The system was thus an environment in which to model interpretation of temporal events and relations—in this case, through graphical means that were linked to structured data outputs. It was a model of interpretation of temporal relations at a higher order, which is the system of modeling itself. Temporal Modeling was meant to support a meta-reflection on the work of interpretation while also offering the means to model interpretatively.

ANOTHER ALTERNATIVE: THE IVANHOE GAME (AND ITS SUCCESSOR, I.NTERPRET)

While the Temporal Modeling project had humanistic engagements with time, temporality, and chronologies at its core, the Ivanhoe Game was designed to support interpretative activity. In this instance, the conviction was straightforward: interpretation arises from engagement with documents, data, information, and speculation on the ways connections among these create a semantic field. The fundamental engagement with textual interpretation formed the foundation of the design of the Ivanhoe Game (Figures 11.3–11.5).[27]

Created as a "game of interpretation," Ivanhoe was designed to support the work of multiple players, each of whom had defined a role within which he or she was working on a shared textual object or corpus.[28] As each player engaged with the text (generally a literary artifact, though nothing in the structure or design of Ivanhoe prevented texts of other kinds from being the focus), their interventions, assessments, additions, changes, or other actions were registered. The game field reflected the focus of attention of each player, but also the shifting aggregate of moves being made in the game as a whole. Because the project had been designed with a point-of-view system, the field of play was always viewed from someone's perspective (built into the design as a filter-governing display). This located the interpretative activity within specific subject positions—that of individual players. As noted, this has obvious resonance with reader-response work like that of Wolfgang Iser or Stanley Fish, but with the difference that the game provided a way to foreground the constitutive interactions in a graphical interface. The game was never seen from "outside" but always from some perspective.

Interpretation emerged as an effect of the sum of interventions in the discourse field and literally took shape within the graphical display mapping the game. The semantics of this display were not legible as a textual analysis (in the way a topic model might be, for instance) because it was not rendered in language but in graphics. But the form and shape of game dynamics and the sites of greater attention within the text under investigation, and additional texts put into the game, could be grasped in the visual organization and arrangement of components on the screen. The premise was that interpretation was what the game was about.

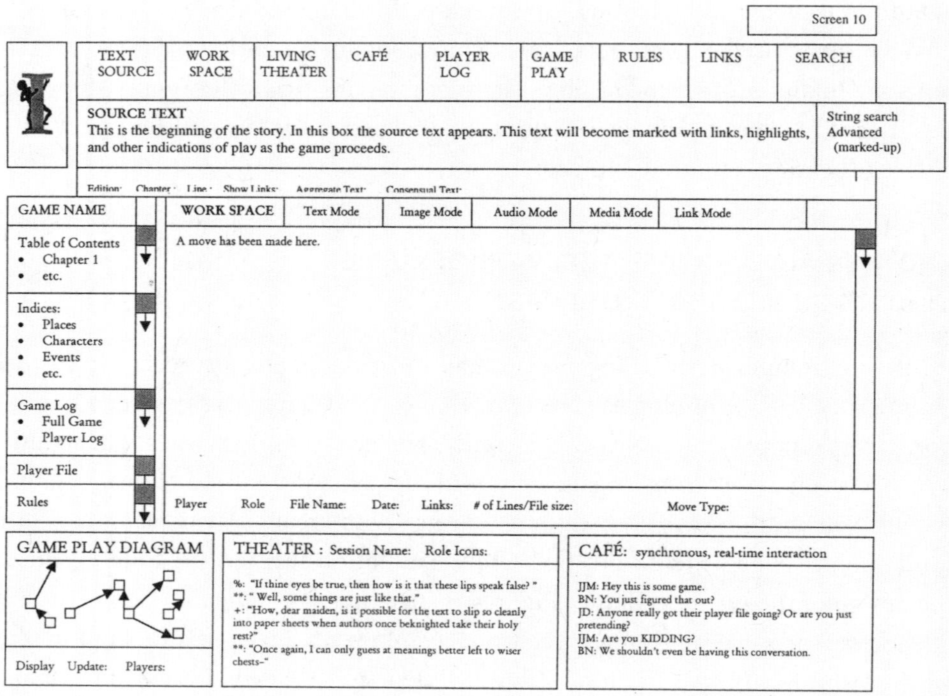

FIGURE 11.3. Ivanhoe—conceptual sketch of activities supported by the game. Drawing by the author.

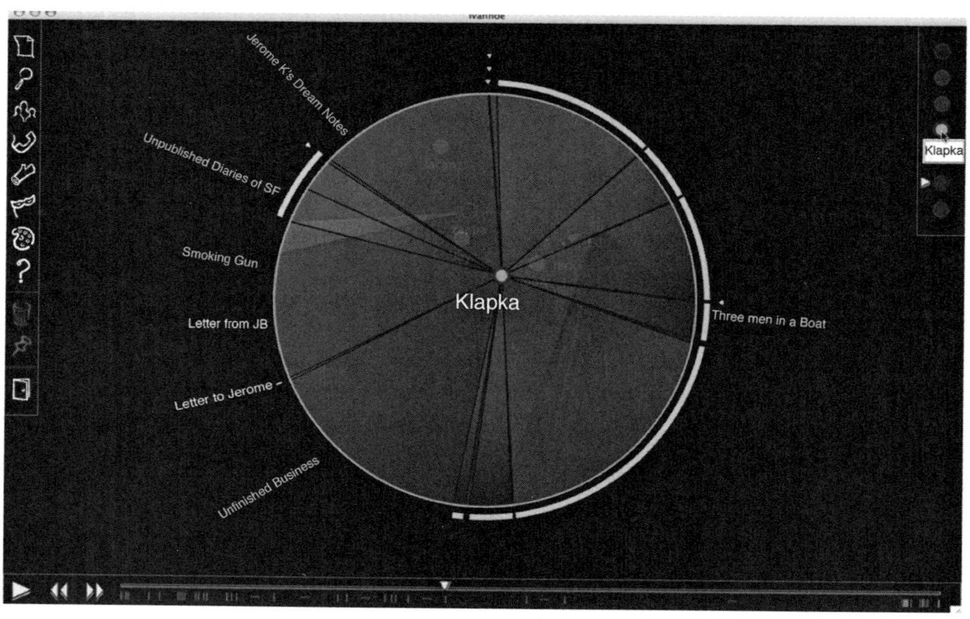

FIGURE 11.4. Ivanhoe—implementation screen of the original. Screenshot from SpecLab team design.

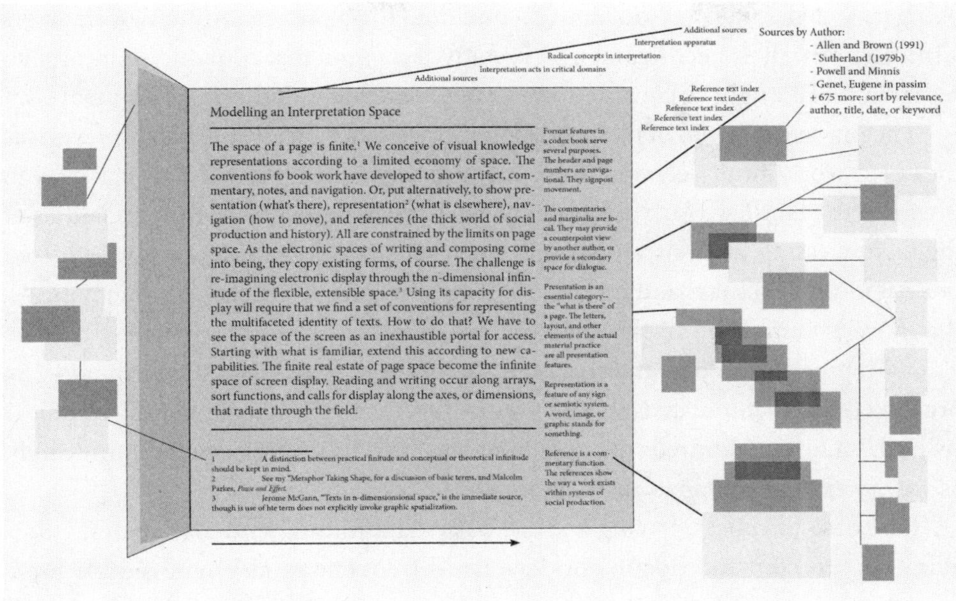

FIGURE 11.5. Document-based I.nterpret workspace. Drawing by the author.

Providing a way for interpretation to be foregrounded as a set of moves or actions was the primary motivation in its design. By showing the shape of argument, and offering components of a system to do the work of interpretation, these two projects demonstrated the value of reifying interpretation and individual acts of interpretative work.

Working prototypes for Ivanhoe and Temporal Modeling were developed within what became SpecLab, the Speculative Computing Lab at the University of Virginia, between 2000 and 2008.[29] Further extensions of these projects appeared in a project proposal I designed, named I.nterpret, that was meant to be a site with greater capacity and flexibility for interpretative work than that supported by Ivanhoe. Whereas Ivanhoe sustained collaborative work focused mainly on one text, I.nterpret was intended to integrate the activities of scholarly research—bibliographical trails and links, text analysis, visualization, multiple (many) documents, and the layering of interpretative frameworks onto the underlying evidence—and vice versa. Based on the premise that interpretation is not merely a restatement of a preexisting work but a constitutive act that produces the text through a reading—of one textual artifact or across many—I.nterpret focused on the virtual space between reader/actant and texts as the site in which interpretation was produced. The complexity of this model, which went beyond the reader–text, interpreter–temporality structures of Ivanhoe and Temporal Modeling, introduced a more explicit recognition of the role of enunciation and subject positionality into the way interpretation was being modeled. I.nterpret took into account that the basic organization of elements within a research project was an interpretative act—and that configurations of such research work ought to be able to be noted, recorded, and subject to analysis.[30]

The implementation of platforms for interpretation in all SpecLab projects was meant to work intuitively and allow for free-form engagement. The idea was to integrate the familiar work habits of analog scholarship with the functionalities of digital processing, analysis, and display. So, in a typical humanities research project, one has multiple sources and resources—texts, images, notes, maps, commentaries—that have accumulated in the course of the research. The first challenge is to be able to represent these in an on-screen surrogate in a way that allows them to be organized and manipulated. Should they be represented by icons? By full-text files? By both? If I have hundreds of pages of notes, I need a way to organize and search, as well as extract and analyze, their contents.

For these purposes, having a set of basic data mining and text analysis tools allows me to compare documents, see word frequencies, do some simple topic modeling (a statistical analysis of frequency and probability relations among thematic terms) and some keyword-in-context visualizations, and even network analysis of the citations I have collected. These digital augmentations need to be able to be integrated into a scholarly argument. So if I begin writing, I might want to (1) reference source materials in a sidebar, (2) show a digital analysis in an inset window, (3) organize my evidence in a graphical interface that lets me show how sources and commentary are related, and (4) preserve this material for publication and iterative rework. I would need some kind of way to store, classify, search, and manipulate these materials within the digital platform without being overwhelmed. A navigable graphical space allows me to model the acts of interpretation through direct manipulation and work.

Suppose I might want to push my analyses into more speculative realms: what are the models of temporality that emerge *from* my documents, for instance, especially if they come from a wide range of historical moments (pregeological timescales, for instance, or biblical ones)? Likewise, I might want to produce a model of a represented space—a domestic or village environment from a study of eighteenth-century novels, or an imaginary geography from the fictive travels of John Mandeville, or the socially charged space of a slave uprising. These geographies and temporalities need to be generated, not represented, because they do not replicate preexisting conventions or necessarily use standard metrics. All of these features are integral to the notion of modeling interpretation.

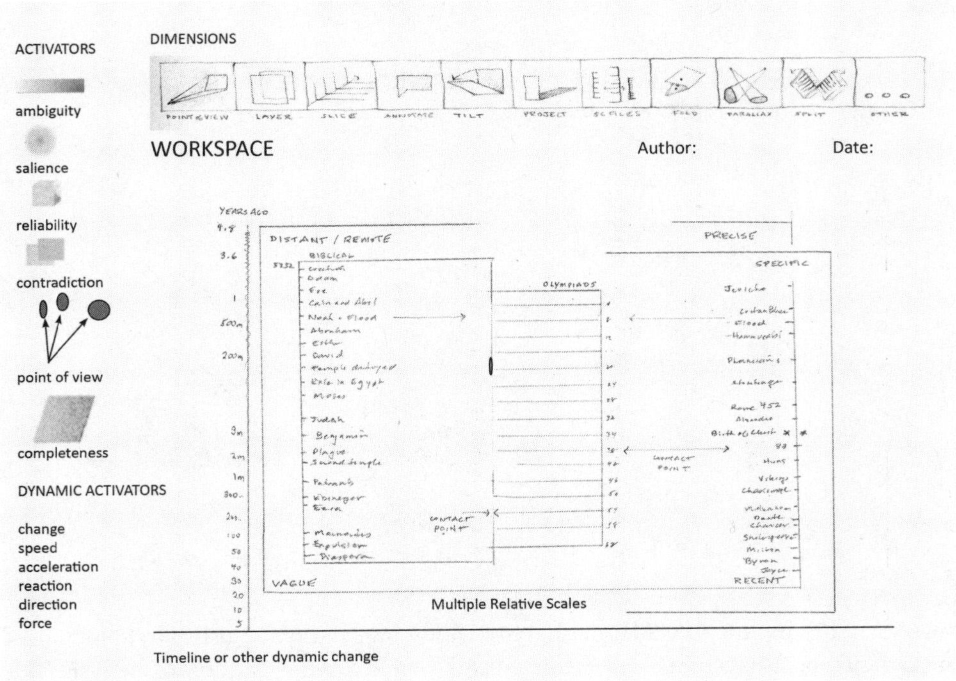

FIGURE 11.6. The 3DH workspace design showing an image of various chronological scales, with dimensions and dynamic activator panels in view. Argument space for arranging evidence is not shown.

CONTINUING WORK: 3DH (THREE-DIMENSIONAL HUMANITIES)

In 2016, the 3DH project, convened by Jan Christoph Meister at the University of Hamburg had the goal of creating next-generation visualization tools for the humanities (Figure 11.6).[31] This project offered the possibility of instantiating an updated approach to nonrepresentational visualization through engagement with the already established narratological work Meister and his team had done in creating a platform for markup, CATMA. CATMA is a highly sophisticated environment for creating XML markup schemes, formalized interpretative frameworks codified in hierarchical tag sets.[32] Ivanhoe had required no technical background or special training (the prototype was used in classroom teaching), and Temporal Modeling was designed as an easy-to-use authoring environment, a graphical canvas. CATMA relies on more specialized familiarity with XML. A user creates a highly customized set of tags for marking up a text in accord with an interpretative scheme but must work within the formal constraints of that XML structure. CATMA could augment Ivanhoe, but it was designed to do a particular kind of narratological work analyzing themes specific to a scholar's engagement with a text.

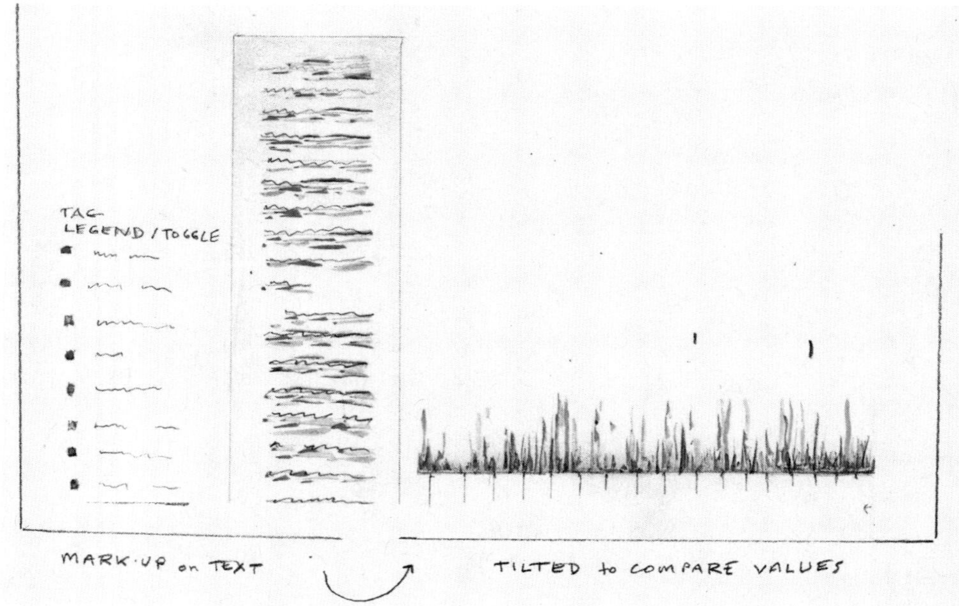

FIGURE 11.7. Sample of CATMA markup showing a tag set (distinguished by colors) used in markup (vertical column) and extracted and rotated to show patterns of analysis. This would appear in the workspace/canvas above.

The 3DH environment was meant to integrate back-end interpretative markup and front-end display with a graphical canvas for engaging in interpretation that would feed back into the database in a structured form.

Like the other platforms and tools already described, the markup platform of CATMA is discipline agnostic. Nothing restricts its use to literary work, and its tool set could be applied just as readily to law texts, medical documents, news transcripts, or financial records as to aesthetic works. CATMA's main function was to allow a reader/user to create a customized XML schema—a set of markup tags in a nested hierarchical structure—that could be applied in the act of reading. These tags function exactly the way multiple-colored highlighters work in an analog environment. They are a way of inserting interpretation through direct graphical input, asserting that this phrase or that word or a paragraph or name should be understood as belonging to a particular theme, topic, mood, or character—or any other category with which a particular tag was associated.

CATMA tag sets can be very extensive and generate a thick field of interpretative output. Seeing the patterns generated by the action of reading was not easy—multiple overlapping tags and heavy markup layered into a text required new discovery and display tools for viewing and analysis (Figure 11.7). Imagine marking up Wolfgang Goethe's *Elective Affinities* with a set of tags that allow the critic to interpret the

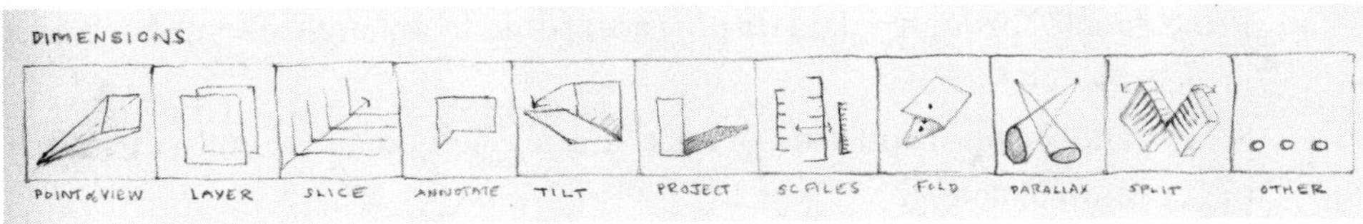

FIGURE 11.8. Explanation of the functionalities of the dimensions in 3DH.

language of the novel across a range of scientific metaphors and tropes of chemistry that align these themes with a model of human passions. The amount of markup and range of terms could become highly elaborate very quickly. The 3DH project focused in part on creating a way to visualize this substantive and extensive work of interpretation in a meaningful way. In this instance, sketches and proposals for prototypes were produced, along with other features of a more extensive graphical environment for manipulating structured data through visual interface and display. The notion of enunciation (of analyzing language in terms of subject positionality) played a very small role in these designs, however, and the notion of *point of view,* a feature of both Temporal Modeling and Ivanhoe, did not (unfortunately) become an integral component of the design vision for 3DH.

In Temporal Modeling, the graphical space was conceived as a primary site of production of interpretation—not display of it. In working on 3DH, I again emphasized the idea of using graphics as the means of creating interpretation alongside the tools of display and discovery that are more familiar features of information visualization. Some proposed features of 3DH included possibilities for "painting" inflections onto node-edge diagrams used in network analysis; manipulating graphical components of charts, graphs, timelines, or other visualizations; and inserting point-of-view systems within data displays (Figures 11.8 and 11.9). In practice, this means manipulating the functionalities of the dimensions in 3DH using the following factors:

- Point of view: assigns an author attribute, creates a now-slider to put multiple values in contrast, uses vanishing points and perspectival systems as appropriate.
- Layer: serves as a marker of partial knowledge and evidence and allows toggles to be activated along a slider to display continuous or discontinuous evidence/argument/narrative.
- Slice: creates a display on attributes of data, tracks patterns of co-occurrence and discrepancy.
- Annotate: allows attributes to be added to data, nodes, edges, texts, images, cells/rows, or elements.
- Title: skews the display along a line of bias or inquiry.

- Project: cuts through a data display according to an angle of emphasis and projects a new display.
- Scales: allows for production of relative scales kept distinct but with points of common exchange.
- Fold: matches data points and patterns to see discrepancies and alignments.
- Parallax: creates multiple view lines into data display, time models, maps, and markup.
- Split: cuts in any dimension to view, slice, see into, move, or change scale of granularity.
- Other: supports shifts of elements to see into, order, arrange, or sequence graphic elements.

The creation of data from direct manipulation of graphical features allows visualizations to be instruments of interpretation, not merely displays of already-processed interpretative work. The concept of a two-way screen was crucial, but so was the development of a graphical vocabulary for modeling interpretation. The components of this argument space include such basics as ordering, sequencing, proximity/distance hierarchies, and other graphical features with semantic or syntactic value.

INTERPRETIVE INTERFACE, MEDIAL IDEOLOGY, AND ENUNCIATION

Each of these projects, Temporal Modeling, Ivanhoe, I.nterpret, and 3DH, contains demonstrations of certain theoretical principles. But to fully understand interpretation as constitutive and then model its rules and procedures, the interface has to be conceived as a space of action. So, in coming to the end of this argument, I will expand the discussion to a larger theoretical framework within which discussions of interface design for interpretative work intersect with concepts of medial ideology, agency, and theories of enunciation.

Issues of agency—the notion that we are doing interpretative work as fully autonomous subjects—intersect with those of medial ideology as described by Matthew Kirschenbaum. This concept addresses the set of beliefs that accompany the use of screens and media to conceal their workings while pretending to expose them.[33] In that process, delusions of agency are produced and reinforced. The user experience modeled on consumer satisfaction offers an illusion of control sustained by satisfaction in the small clicks, snaps, swipes, and other feedback built into the system.[34] Every action has a reward of some kind, and this produces a sense of achievement, accomplishment. Shifted into another critical frame, we can approach this situation by offering an alternative description of interface as a

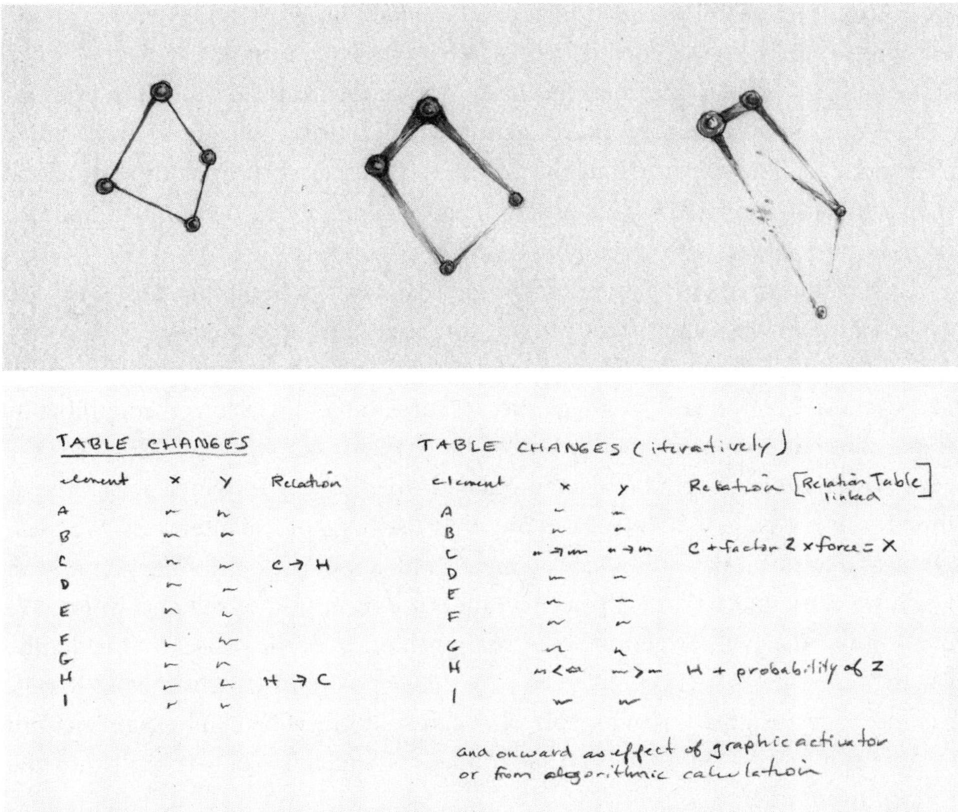

FIGURE 11.9. Example of 3DH node modification through graphic manipulation and a table with remodeled data.

system of enunciation that produces both a speaking and a spoken subject and, in the process, an illusion/delusion of agency. Standard interface theory, particularly within the engineering approach of the human–computer interface community, assumes an autonomous, free-willed individual subject—one in full rational control of agency and action. By introducing concepts of enunciation, the goal was to demonstrate that subject positions were part of dynamic systems of power relations, not autonomous agents. One of the research questions asked how the concept of linguistic shifters, such as those identified by Roman Jakobson and discussed by Émile Benveniste and others, could be used to analyze graphical features.

The concept of enunciation invoked here arises from structural linguistics, in particular the work of Benveniste on pronouns and other shifters, terms whose value (such as *I, you, here,* and *now*) depends on their locutionary setting to be filled out with semantic reference.[35] By approaching interpretation within the framework of language as an enunciative system, the roles of speaker and spoken subject are called to attention. The enunciative apparatus also makes explicit the social condition of any articulation and its embeddedness within systems of power relations and ideological constructions. When we extend these linguistic notions to the analysis of interface, then the issue of the constitutive model of interpretation becomes involved in a scene of action where illusions of agency and control are produced. The presumed agency of an individual working with a graphical interface is also modeled by that environment—generally, in current use, a screen display that serves as a front end to computational activity.[36]

Critical theory, once again, seems to get jettisoned in the implementation and use of digital environments that provide a sense of omnipotence—or, at the very least, potency. The ways this (illusion) is structured into the graphical environment has parallels with the way language use obscures subject relations through habits of use, familiar constructions, and other naturalizing activities. We don't see the ways power is constructed and manipulated—or the way in which we are positioned within systems of control—in our daily use of graphical interfaces. We click, swipe, and follow links without stopping to think about the disciplinary actions to which we are subjected and their effect on our conception of self and world. We might notice the overt targeting of our user profile, but do we pause to consider the way the entire construction of screen experience positions us as subjects? Rarely. The interface design always controls what can be done, what can be "said" within its constraints, and thus the way its "model" of subjects—us—is built into the possibilities it offers.

Miriam Posner has been researching supply chain software and its perverse combination of ubiquitous viewpoints (global maps providing an overview of every point in the known system) and complete opacity (the concealing of all decisions, actions, costs, and impacts of keeping the supply chain working).[37] The model instantiated here is one of control over events and decision processes; the user is interpreting the data flows mapped on the screen but has no input into the processes generating the data. As in the standard graphical user interface at work on screens of all kinds—computer terminals, mobile devices, tablets, and wall displays—the system creates a model of agency according to a consumer version of individual identity. Ease of use, satisfaction from behavior of the components, and other features of the system are designed to satisfy a user as immediately as possible, providing gratification in the form of haptic, auditory, visual signals. This reinforces an imagined agency conceived as ubiquitous—or at least omniscient—with regard to the system. And yet, the very concept of enunciation bears within it a recognition of the located, partial, and specific properties of individual position within any social system. Modeling the shape of enunciation in interface is a radically different task than simply modeling behaviors of components or actions of a user. The sociality of enunciative actions suggests the need to model agency within terms of accountability and responsibility.

In some sense, every engagement with an interface—whether analog or physical, such as printed texts; theater stages; visual art, digital and virtual; screens; and immersive environments—is part of an enunciative system and part of an environment that sustains and provokes interpretation. Interpretation in the work of Temporal Modeling, Ivanhoe, and 3DH is both the focus of the designed environments and also a basic critical framework for the practices in which interfaces are involved. The pervasiveness of interpretative acts (as a fundamental feature of epistemology viewed from a constructivist perspective) makes an argument for knowledge as a modeled effect of interpretative acts. The role of graphical, visual, and now digital means is in part to expose the model of these acts and of the systems and conditions within which they arise. Interface provides the appearance of agency—the ability to move, manipulate, or create within its parameters. To reiterate what was stated earlier, in a profound sense, the interface models what is possible; what can be said and how; what can be spoken, graphed, visualized, and positioned within the system.

In conclusion, we can see that the concepts articulated in relation to modeling interpretation in a digital environment are not new, nor are they unique to the media of networked or electronic platforms. Interpretation is understood throughout this study as a primary moment of engagement between interlocutor and phenomenon—whether this is constituted by a reader/text, researcher/situation, or algorithm/corpus. But under no circumstance is the interpretative act seen as a mere uptake or transfer of the text, event, or materials into an inscribed, written/graphical/numerical expression or mental, cognitive impression. Interpretation is a reified outcome of a dynamic process, a third component to the system of intellectual (or computational) exchange, and it is, of course, never the same as the provocative source from which it arises. Every act of interpretation is observer dependent, not disengaged or objective—although it is also objectified, reified. Visual systems model enunciation as surely as verbal ones, and the structuring features of these systems are already models of their assumptions about what interpretation might be.

Making the models by which interpretation works explicit, calling attention to the sociality of enunciative actions, stressing the accountability and responsibility of enunciative acts as constitutive acts, puts them back into a hermeneutic framework where every feature of the system can be described explicitly. Thus the basic concept of agency and model of its enactment can be modeled as a study of interpretative actions and constitutive acts. That these are based in models becomes more clear as we struggle to design environments that make the rules of the modeling more explicit. This does not release us from the inevitability of disciplinary regimes of enunciative systems but merely gives us the tools to understand their workings and the constraints on our models of agency and illusions thereof. The very frames by which we enclose the environments we design for our work are the boundary limits of what the model permits, no matter how aspirational its rhetoric and goals or how open-ended its capacity for expression.

NOTES

1 Willard McCarty, "Modeling: A Study in Words and Meanings," in *A Companion to Digital Humanities,* ed. Ray Siemens and John Unsworth, 254–72 (Oxford: Blackwell, 2004). McCarty is generally cited within digital humanities for his discussion of modeling, though I will note that our Temporal Modeling project was conceived in 2000, several years before this article. http://digitalhumanities.org:3030/companion/.

2 Nigel Thrift, *Non-representational Theory: Space, Politics, Affect* (New York: Routledge, 2007).

3 Plenty of work on the effects of algorithms and their biases exists, but it does not address the actual encoding and scripting to read the model that produces these outcomes.

4 Johanna Drucker, *SpecLab: Digital Aesthetics and Projects in Speculative Computing* (Chicago: University of Chicago Press, 2009).

5 Gerald Bruns, *Hermeneutics: Ancient and Modern.* (New Haven, Conn.: Yale University Press, 1992).

6 Bruns, introduction to *Hermeneutics.*

7 Paraphrasing Bruns throughout this paragraph and his discussion of the "revolution" in twentieth-century hermeneutics.

8 Jerome McGann, *Radiant Texuality* (London: Palgrave Macmillan, 2001).

9 See my report on the work done in the context of 3DH. Johanna Drucker, "Non-representational Approaches to Modelling Interpretation in a Graphical Environment," *Digital Scholarship in the Humanities* 33, no. 2 (2018): 248–63.

10 Thrift, *Non-representational Theory.*

11 Johanna Drucker, "Humanities Approaches to Graphical Display," *DHQ: Digital Humanities Quarterly* 5, no. 1 (2011), http://www.digitalhumanities.org/dhq/vol/5/1/000091/000091.html, and also Drucker, "Humanities Approaches to Interface Theory," *Culture Machine* 12 (2011), http://www.culturemachine.net/index.php/cm/issue/current.

12 Johanna Drucker, *Graphesis* (Cambridge, Mass.: Harvard University Press, 2014). See also Peter Galison, *Image and Logic* (Chicago: University of Chicago Press, 1997).

13 Michael Friendly, *Milestones in the History of Data Visualization* (New York:

Springer, 2005); Edward Tufte, *The Visual Display of Quantitative Information* (Cheshire, Conn.: Graphics Press, 1983). William Playfair is justly renowned for his innovations in making charts that display quantitative information. His accomplishments were to make patterns in numerical data legible through graphic patterns and to make these visualizations aesthetically engaging. This chart contrasts the value wages (by week) and price of wheat (tracked by decade) against a grid bracketed by the reigns of British monarchs. The tints, tones, delicate line work, and engraved lines are elegant features of their rhetoric, the impact of which is carried, in part, by these qualities. Argument rather than accuracy and rhetoric rather than logic characterize his graphics, even though they are universally seen as rationalized presentations of information. In this often-reprinted image, for instance, we see that the periodization of data gathering is different for wheat prices than it is for wages, but they are presented in parallel, obscuring this decision. His work builds on practices identified by the British statisticians John Graunt and William Petty, termed *political arithmetic*, a phrase that makes clear the connection between social management and information that becomes integral to the modern era.

14 Edwin W. Kopf, "Florence Nightingale: The Lady with the Data," March 15, 2016. https://thisisstatistics.org/florence-nightingale-the-lady-with-the-data/.

15 W. E. B. Du Bois, http://www.webdubois.org/wdb-1900exp.html.

16 Drucker, *SpecLab.*

17 An early white paper, with coauthor Bethany Nowviskie, is available at http://www2.iath.virginia.edu/time/reports/infodesign.doc.

18 This image combines features (top bar, inflections, bottom slider) from the original Temporal Modeling project and an updated vision of stretchy timelines and data views. The original Temporal Modeling proof of concept (2001–3) succeeded in showing that a graphical workspace could be used to generate structured data within a humanities context. Despite certain limitations (in particular, the standard metric embodied in the screen display), the project embodied a few key features: a now slider, a set of syntactic and semantic inflections, a way to create multiple timescales and layers, and the ability to transform temporal models over the course of unfolding events. The functioning prototype did generate structured data in XML format that could be exported and used for manipulation, modeling, or analysis. Design features like "stretchy timelines" or "multiple chronologies" or even causal relations among temporal elements were not implemented after the initial conceptual design phase and are shown here in an updated vision of the project.

19 James T. Fraser, *Time, the Familiar Stranger* (Amherst: University of Massachusetts Press, 1987), and Fabio A. Schreiber, "Is Time a Real Time? An Overview of Time Ontology in Informatics," in *Real Time Computing,* ed. W. A. Halang and A. D. Stoyenko, 283–307 (Berlin: Springer/NATO ASI, 1992).

20 In Einstein's description of space-time, now is to time as here is to space—a location in a topographic field. Time is spatial rather than directional, and points in time relate to each other as points on a map do; they are not organized on a line. For a clear, popular description of this concept, see Carlo Rovelli, *The Order of Time* (New York: Riverhead Books, 2018).

21 Standard references on time and narrative include Gerard Genette, *Narrative Discourse: An Essay in Method,* trans. Jane E. Lewin (Ithaca, N.Y.: Cornell University Press, 1980); Seymour Chatman, *Story and Discourse* (Ithaca, N.Y.: Cornell University Press, 1978); and the monumental work by Paul Ricoeur, *Time and Narrative,* trans. Kathleen McLaughlin and David Pellauer (Chicago: University of Chicago Press, 1984).

22 Jeremy Norman, "Eusebius Introduces His Tabular Timeline System," *History of Information* (blog), http://www.historyofinformation.com/expanded.php?id=3804.

23 James Ussher, *Annales Veteris Tesamenti* (Dublin, 1650).

24 Sir Walter Raleigh, *History of the World* (London: Walter Burre, 1612); Sir Isaac Newton, *The Chronology of Antient Kingdoms Amended* (London: Tonson, Osborn, and Longman, 1728).

25 A monumental example is the sixty-five-volume project by George Sale et al., *An Universal History, from the Earliest Account of Time* (London: T. Osborne, 1747), but work by Denis Petau (1659), William Whiston (1708), and many others could be cited. See also Anthony Grafton's work on Joseph Scaliger, *Historical Chronology,* vol. 2 (Oxford: Clarendon Press, 1983).

26 Work on this phase of the project was a collaboration with Bethany Nowviskie.

27 Conceived as a multiplayer game of interpretation that would allow a group of users to perform textual analysis through manipulation, commentary, and role-playing. The conceptual design contained a palette for making "moves" in the game and was envisioned as a workspace for scholarship, mainly in text-based projects, as is suggested by the activity shown. The actual implementation deviated from the original concept in certain ways. It provided an interface into the back-end features, such as the "log" that tracked each transformation in accord with a "garden-of-forking paths" metaphor. But the game space featured the entire "discourse field" displayed from the point of view of a particular player. The perimeter of the circle indicated texts and documents, while the lines of connection indicated the course of play as a set of exchanges among players. Lost in this implementation was the legible display of documents and materials brought into play—or the ability to manipulate them as a set of research materials.

28 Johanna Drucker, "Designing Ivanhoe," *Text Technology* 12, no. 2 (2003): 19–41.

29 Drucker, *SpecLab.*

30 This project was never realized, and only design specifications were created, not an implementation.

31 Jan Christoph Meister, Geoffrey Rockwell, Rabea Kleymann, Evelyn Guis, Marco

Petris, and others (Marian Dork and Jan-Erik Stange came into the project more recently). See documentation at http://threedh.net/. Designed to create structured data as well as to modify them at the level of display, 3DH was meant to provide a space for working with humanities projects at various scales. The project made a distinction between the spaces of display/analysis and production/interpretation. In the first, the display was based on preexisting data; in the second, the primary goal, as in the earlier projects, was to be able to work directly with a set of graphical tools that could provide interpretation. Some of these were deliberately speculative, such as the tools for cutting, slicing, tilting, or performing other graphical manipulations on an initial display. These took as their point of departure the conviction that principles of analysis from geometry (such as projection onto a plane) could be used to investigate data through visual means. Other features, such as the ability to add inflections (importance, change of values over time, and so on), were meant to work as an interpretative "paint box" to add dimensions to the data display that would become part of the underlying data—or even change the data structure.

32 See http://catma.de/.

33 See the chapter "Every Contact Leaves a Trace" in Matthew Kirschenbaum's *Mechanisms* (Cambridge, Mass.: MIT Press, 2007).

34 Conventional approaches to interface design are based on principles of reward, cognitive load, and other tenets that assume a consumer model of the user rather than a subject. Ben Shneiderman's "Eight Golden Rules of Interface Design" is a well-known reference in this regard. https://faculty.washington.edu/jtenenbg/courses/360/f04/sessions/schneidermanGoldenRules.html.

35 See Johanna Drucker, "Information Visualization and/as Enunciation," *Journal of Documentation* 73, no. 5 (2017): 903–16.

36 Alex Galloway, *The Interface Effect* (Cambridge: Polity Press, 2012), provides a critical approach; Ben Shneiderman, *Designing the User Interface* (Reading, Mass.: Addison-Wesley, 1987), is the standard reference for the engineering approach.

37 Miriam Posner, "See No Evil," https://logicmag.io/04-see-no-evil/.

Afterword
On the Humility of Models

SARAH WASSERMAN

Bringing material culture studies to bear on the expansive intellectual terrain of modeling, as this volume has done, testifies to the central role that modeling plays in the humanities. The essays gathered here challenge the commonplace assumption that modeling is the work of scientists and engineers. Instead, the volume shows how much modeling matters to historians looking to tell the stories about those scientists and engineers and to art historians, philosophers, musicologists, and literary critics. To approach modeling as a fundamentally humanistic pursuit is to see that it is not only about finding meaning but about making it. The term *modelwork* names the emphasis that material culture studies places on the components and techniques that bring a physical object into being. Models are not merely derivative translations of phenomena and entities that exist in the world; they must be thought up, drawn, designed, assembled, sculpted, scaled, calculated, programmed, tested, refined, and revised. The sum of these activities and the way they in turn act on the people who make them—the *work* part of *modelwork*—is what the essays in this volume make visible and material, from the shoe last to digital models of time. In shifting the discourse of modeling, they also reflect back onto the field of material culture studies, reminding scholars interested in artifacts that we should do much more than study finished products. Modeling—the gerund form pointing to its ongoing, iterative nature—stresses the way that objects and ideas are both works in progress.

This approach is intuitive to anyone who has built a model: the toy ship, for instance, enthralls because it is an elaborate miniature, but also because it requires patience, attention, and time to assemble. Despite the commonplace understanding of models as things that are made through dynamic processes, this fact is often overlooked by scholars. It was necessary to assemble this volume because while models are often predictive and prognostic, the term tends to signal something concrete and empirical: on one hand, *theory*, with its recursive questions and destabilizing abstractions, and on the other, *models*, with their reassuring relationship to the real world. Such a binary doesn't hold for long: theories are models about the world and the subjects who inhabit it; models are theories of how something exists and operates. So while we often think of models as tools for standardization—devices, plans, and equations that ensure the consistency and repeatability of designs and experiments—the essays in this volume stress models' protean nature. They help us see that whether we are talking about black holes or Byzantine churches, models require interpretation. Even as models themselves translate objects and forces into drawings, algorithms, and miniatures, they necessitate translation by a sensitive user or reader. Rather than establishing a single, empirical truth, models—especially those discussed in this volume—open onto many possible meanings. They thus invite the kinds of scholarly labor that are the hallmark of the humanities: historical research, linear narration and thick description, context building, and interpretation.

I will return to the centrality of work and its relationship to materialisms both old and new in a bit, but first I want to linger with the issue of interpretation. The curious status of the model as both mimetic and prophetic raises questions about how it relates to contemporary trends in literary and cultural studies. If the model is humble—an approximation or copy of something else—then it might be said to align neatly with the current emphasis on descriptive methods. Bruno Latour's actor–network theory, Rita Felski's notion of postcritique, and Paul Saint-Amour's concept of weak theory all name interpretive modes that eschew certitude in favor of a more modest orientation to our objects of study.[1] But if we understand models as ideals that displace the objects or systems they represent, they might have more in common with the recent turn in the humanities toward making and toward data. Now that some literary scholars can write code and the computational humanities can "read" thousands of texts, it appears we have new, or at least different, mastery over our subject matter. Perhaps it is precisely because models are simultaneously emblems of humility and hubris that they have risen to new levels of importance and interest across disciplines.

The paradoxical quality of models that makes them seem both modest and masterful has generated vigorous debate about the digital humanities (DH) and whether this relatively new field of study makes unjustified claims toward empirical certainty. In 2013, Johanna Drucker offered a simple but capacious definition of the burgeoning field: "Digital humanities is work at the intersection of digital technology and humanities disciplines."[2] DH work assumes many different forms—it is as varied as, say, scholarship on the early modern period, which can be literary historical, highly theoretical, or focused on visual artifacts or even aural ones. In the second volume of *Debates in the Digital Humanities* (2016), Lauren Klein and

Matthew Gold acknowledge that this range of approaches makes it difficult to define the field:

> Along with the digital archives, quantitative analyses, and tool-building projects that once characterized the field, DH now encompasses a wide range of methods and practices: visualizations of large image sets, 3D modeling of historical artifacts, "born digital" dissertations, hashtag activism and the analysis thereof, alternate reality games, mobile makerspaces, and more. In what has been called "big tent" DH, it can at times be difficult to determine with any specificity what, precisely, digital humanities work entails.[3]

Despite the heterogeneity of DH scholarship, much of it in fact uses models and modeling to advance its claims. Whether these models are designed to test (as in a prototype digital game used to scaffold later iterations) or predict (as in the case of algorithms designed to analyze text), DH has long invested in modeling as one of its primary methodologies.

Can DH, with its robust discourse about models, offer clues about what modeling holds in store for the humanities? In *Distant Horizons* (2019), Ted Underwood uses statistical modeling to argue that the historical periods scholars conventionally use to separate literature have foreclosed knowledge about continuities of theme, style, and genre across periods, blocking our view of the long arc of literary history. Underwood defines modeling simply as "a relation between variables."[4] To rethink period and genre, Underwood draws upon logistic regression, a form of predictive analysis that explains the relationship between a dependent variable and one or more independent variables (in Underwood's study, for example, the category of fiction as distinct from biography, based on different verbs of sensory perception). This kind of modeling aims not just to verify claims with data but to run multiple sets of trial-and-error experiments, leading to an iterative process of guessing, testing, and refining, rather than to tidy answers. As Dan Sinykin says of *Distant Horizons,* "against the purported objectivity of algorithms, [Underwood] leverages the human prejudices built into modeling toward humanistic ends."[5] Sinykin and Underwood thus make the case that models are unlike *theory* insofar as they refuse claims to theoretical universalism and objectivity. In this reading, theory is "a system of ideas intended to explain something, especially one based on general principles independent of the thing to be explained," whereas modeling always remains messily tied to its particular objects.[6]

Feminist DH scholars, including Katherine Bode, Catherine D'Ignazio, Lauren Klein, Tara McPherson, and Whitney Trettien, have demonstrated how the data sets, visualizations, and software on which models rely are generative of knowledge precisely because they are imperfect and idiosyncratic.[7] In *Data Feminism,* D'Ignazio and Klein forcefully illustrate how frequently data and the supposedly neutral methods used to collect, compile, and communicate them are gendered.[8] Bringing to bear the lessons of theory (in this case, feminist theory) on DH, such scholarship reveals how important it is to account for bias and look for ways to move beyond it when engaging with models. Recent work by Andrew Piper offers additional insight into the relationship between theory and modeling. In *Enumerations: Data and Literary Studies* (2018), Piper claims that using computational modeling to analyze literary texts forces scholars to reflect on their own assumptions. Because they must test their claims against the "external" mechanism of the model, such scholars reflect explicitly on their intuitions and interpretations, laying them bare for other readers to see. Piper's argument is that modeling "run[s] strongly counter to much of the language of empiricism that has surrounded the initial rise of the field [DH]. We are always present in our models."[9] According to Piper, modeling stands in relation to DH in a similar way that theory has to "analog humanities." For Piper, modeling enriches the more raw empiricisms of DH by inflecting them with human creativity and interpretive labor. Just as theory has been used to interrogate and complicate historical and formalist literary and art criticism, models circulate within DH as generative disruption because they are propositional and subjective.

This disruptive, creative potential of models comes in large part from what we might think of as their curious lag-lead time. The essays in this volume repeatedly ask a version of the chicken–egg question: which came first, the model or the thing modeled? Would the concept of the "modern human form" exist in quite the same way without the sculptures commissioned by Dudley Allen Sargent? Does the composite, historical site called the Church of Studious really exist outside the hypermodel that architects and archaeologists have constructed to visualize its many layers? Annabel Wharton names this tension when she writes that "all models have referents" but also "a vast range of ways of relating to those referents." This formulation recalls a traditional view about models that sees them as "material analogies," which allow scientists to develop a more easily grasped picture of a complex principle. Material here is used to indicate that such a model usually involved a comparison to something well understood and concrete. Michael Weisberg neatly sums up this view using a familiar example: "the mathematics describing the propagation of light might be accompanied by an analogy comparing the propagation of light waves to the propagation of water waves."[10] Another way to describe this particular trait of modeling, borrowing terms from linguistics and rhetoric, is to suggest that models operate like metaphors and are likewise composed of a tenor and a vehicle.[11] In *The Philosophy of Rhetoric* (1937), I. A. Richards famously describes a metaphor as having these two parts. The tenor is the subject to which attributes are ascribed, and the vehicle is the object whose attributes are borrowed to "drive" or deliver the metaphor.[12] In Shakespeare's famous metaphor "all the world's a stage," "the world" is the tenor and "a stage" is the vehicle.[13] Following this logic, the Royal Navy's HMS *Victory,* docked in Portsmouth, England, is the tenor; a small replica perched inside a glass bottle would be a vehicle. However, that mighty ship at the dock could itself be the vehicle if we imagine that the tenor is the design or blueprint that preexisted the vessel. Tenors and vehicles come in long chains of antecedent objects and ex post facto derivations. They can also switch places with each other.

It is striking that every essay in this volume establishes one or more tenor–vehicle relationships but then goes on to reverse it. Hilary Bryon shows that William Farish's models, built to demonstrate abstract processes, themselves inspired perspectival drawing techniques. In her reading, a new form of abstraction is generated by a concrete model that derives, in turn, from a processual abstraction. Christopher Lukasik's consideration of eighteenth- and nineteenth-century drawing books reveals that the visual world and the trained illustration technique used to represent it became mutually constitutive. Martin Scherzinger shows how the software designed to model music in fact comes to shape the creation of music itself. Peter Galison's work on scientific models of black holes unexpectedly closes the distance between the far reaches of outer space and the gurgling drain of an ordinary tub. Seher Ford's work modeling the Church of Studious demonstrates how the digital model she creates comes to stand in for the site itself, so that future researchers might in fact take her model as their own. Catherine Howe reveals the way that the composite, statistical sculptures—tenors for "average" American men and women—in turn become the "norm" to be modeled. Reed Gochberg's research on patent models explains how models intended to document a given innovation become sites of innovation themselves. Lisa Gitelman's exploration of shoe lasts traces the bidirectional relays between model feet, model shoes, and the forms of industrial production that shoemaking helped shape. Juliet Sperling's investigation of nineteenth-century anatomical books that model the body ultimately revises understandings of the body and the tactile nature of dissection and medical practice. And Johanna Drucker's projects take interpretation as their tenor and look for nonrepresentational forms to serve as vehicles; in doing so, they reveal that all tools are observer-dependent.

The long arc of modeling and its varied life over time and across media in this volume attest to modeling's ontological instability or, more simply, its generative weirdness. Models stand in varied and changing relation to their referents, challenge our assumptions of modeling's mimetic function, and make visible the gap between idealized and actual forms. This last quality is what Richard So diagnoses in his 2017 essay "All Models Are Wrong." What is "wrong" about statistical models, So explains, borrowing the 1976 aphorism from the statistician George E. P. Box, is that models are based on numbers, which "are simplifying and coarse" and "cannot represent the complexity of the social and natural world."[14] Rather than seeing this as a problem, So argues that recognizing *how* models are always wrong invites us to think with and through them as mediating, interpretive figures. This should have pointed consequences for literary scholars, who, according to So, "have long cast a suspicious and critical gaze toward modeling" because "models run counter to the deep and intensive reading that literary critics take pride in, the exposing of nuance and singularity in texts, writers, and human beings."[15] It is indeed the "wrongness" and strangeness of the models discussed in this volume that illuminate the nuance and singularity So identifies as the hallmark of literature. But So perhaps overstates the opposition between modeling and reading—after all, even if models are an appropriately humble heuristic for literary analysis, close reading and looking have long seen themselves as borrowing power and charisma from the glowing aura of the beautiful objects they describe. As the similarities, rather than the polarities, between modeling and reading come to light, humanities scholarship is poised to become even more interdisciplinary in ways this volume anticipates.

The recent emphasis on the modesty and messiness of modeling within DH, combined with its use of computer code and machine reading, dovetails in provocative ways with the posthuman turn across interpretive disciplines. Even as posthumanists argue that models reflect and refract our humanness, they demote us to a scholarly backseat. Gone is the charismatic critic; in his place is a humble technocrat who models. As we witness the ascent of the algorithm, it's worth recalling that the primary lay referent for models is a human body. Ironically, the essays in this volume do not address what is arguably the most common association with modeling—the fashion model—although on the runway and in catalogs, she certainly highlights the distance between the ideal and the norm. The fashion model's appearance is used to display and market everything from clothing and cosmetics to cars and cheeseburgers. The model's "appeal" derives from her conformity to standard measurements as well as ideas of facial symmetry, gender norms, racial conventions, notions of grooming, and so on. The fashion model simultaneously represents a standardized type and a singular beauty. The visual field is thus saturated with images of models doing what this volume shows that models do: shuttling between the real and the ideal, the thing modeled and its imagined form.

In her 2001 novel *Look at Me,* Jennifer Egan darkly captures this tension. Protagonist and Manhattan model Charlotte Swenson undergoes facial reconstructive surgery after a car crash that leaves her attractive but unrecognizable. Egan uses this plot point to reflect on the inherent strangeness of modeling, the alienation she sees as its primary requirement. Reflecting on her healed face, Charlotte narrates,

> I'd postponed its reckoning with the world for the simple reason that I still didn't know what I looked like. I'd spent as long as an hour staring through the ring of chalky light around my bathroom mirror; I'd held up old pictures of myself beside my reflection and tried to compare them. But my sole discovery was that in addition to not knowing what I looked like now, I had never known.[16]

Charlotte realizes that in offering her face to the world as an object to be molded, made over, and consumed, it ceased to belong to her. In becoming a model, Charlotte has lost the tenor that is herself. What happens, Egan's novel asks, when we recognize that models never correspond neatly to their referents? By exposing the ideologies of power at play in the world of fashion models, *Look at Me* reminds us that all models are steeped in ideologies of power. No practice of modeling—on the runway or in the chemistry lab—is ever fully neutral; this is another way in which all models are wrong.

At her first modeling job after her accident, Charlotte arrives at a photo shoot, where she learns that the artistic director intends to cut her newly healed face with a razor blade so that the photos will "get at some kind of truth here, in this phony, sick, ludicrous world."[17] Charlotte is horrified, and as soon as she refuses, another model is ushered in to take her place. "It's your lucky day, honey," the photographer tells Kimmy, a Korean model, who offers her face up to be cut.[18] Blood drips from Kimmy's wounds, but "the Korean girl seemed not to notice. She looked straight ahead, enduring this assault with the incomprehension of one who accepted long ago that suffering has no purpose."[19] Into this single grim scene, Egan compresses the ethical wrongs of modeling—its exploitative, demeaning, sexist structures. She also confronts its damaging entanglements with race by making Charlotte's replacement Korean. Kimmy is model and model minority, quiet and uncomplaining even as violence is done to her. Egan reveals the perversity of the model minority trope: Kimmy's compliance nets her the job, but at great cost. As Ellen Wu chronicles in *The Color of Success,* the ideology of the model minority harms Asian Americans and perpetuates the notion that America affords equal opportunities to racial minorities, obscuring the truth that the nation is built on anti-Black racism and White supremacy.[20]

Egan's novel and its treatment of fashion models stress that modelwork is work and that work is embodied. As my coeditors note in the introduction to this volume, modelwork is not "the 'immaterial labor' of cultural work proposed by Marxist theorists"; instead, it "straddles . . . the material and immaterial realms." Material here, in the context of modeling, means the clay, stone, and plaster of sculptures; the paper of an illustrated book; and the keys and screens of a computer. But it also means the material bodies of the humans who create and use models and the humans who are models and whose skin, blood, and bones compose another set of raw materials. There is, as this volume points out, no model without its modeler. In historical materialist terms, there is no work without the worker. Black, postcolonial, feminist, and queer Marxist thinkers have long shown that a more thorough reckoning with Marx's ideas of labor power entails a recognition of the way that labor itself has always been embodied and therefore racialized, gendered, differently aged and abled.[21] The importance of (re)-embodying Marxist ideas—by digging into Marx's texts or by subjecting them to new forms of scrutiny and inquiry—is succinctly and forcefully articulated by the Combahee River Collective in their 1977 statement that "we need to articulate the real class situation of persons who are not merely raceless, sexless workers but for whom racial and sexual oppression are significant determinants in their working/economic lives."[22] As a resolutely material concept, modelwork provides a lens for foregrounding the embodied histories of modeling. The question models ask, iterated in this volume's introduction, is not only "Can you imagine that . . . ?" but "*Who* can imagine that . . . ?" Under what conditions? And when? The question becomes one of embodiment and of history; it becomes one that scholars who take material culture studies into account are especially poised to answer.

To return to Egan's novel, one crucial reason for Charlotte's unmooring is that the person she has become—at least physically—since her accident seems to delete the person she has been, a person who looked a certain way and a person who might once have leaned into the razor blade. Whoever Charlotte the model was no longer exists and, even more distressing, may have never existed. Egan's novel thus prompts us to consider the temporality of modeling. Human models must exist in a timeless present; as Charlotte fights to rejoin the fashion world, she must confront the fact that she is not only physically altered but also older. It might be said of all models, whether they are digital, mathematical, or material, that they too are presentist, yet subject to time. Modeling entails looking at an object and attempting to represent it visibly and functionally in the eternal present of the model itself. The work of abstraction that is central to the model usually effaces history. The essays in this volume, however, reinvigorate the histories embedded in modeling and modelwork. Like Egan, they tell the "backstory" of models: the culture out of which the shoe last emerged, the genealogy of the perspectival drawing. More accurately, then, these essays don't tell stories; they *retell* them. More than simply describing models or claiming a sterile, empirical stance, the scholars here narrate, explicate, and contextualize models so that readers see what they mean and why they matter. No matter how contemporary and forward looking the model under consideration, the contributors attest to the importance of historical and historicist knowledge. They gesture toward a road ahead in which interdisciplinary scholarship integrates modeling in the present with a close attention to what has come before. The future of modeling, as it turns out, lies at least partially in the material record of its past.

NOTES

1 See Bruno Latour, *Reassembling the Social: An Introduction to Actor-Network-Theory* (New York: Oxford University Press, 2005); Rita Felski, *The Limits of Critique* (Chicago: University of Chicago Press, 2015); Paul Saint-Amour, "Weak Theory, Weak Modernism," *Modernism/Modernity* 25, no. 3 (2018): 437–59.

2 Johanna Drucker, introduction to *Intro to Digital Humanities,* UCLA Center for Digital Humanities, http://dh101.humanities.ucla.edu/?page_id=13.

3 Lauren F. Klein and Matthew K. Gold, "Digital Humanities: The Expanded Field," Debates in the Digital Humanities, http://dhdebates.gc.cuny.edu.

4 Ted Underwood, *Distant Horizons: Digital Evidence and Literary Change* (Chicago: University of Chicago Press, 2019), 19.

5 Dan Sinykin, "Distant Reading and Literary Knowledge," Cultural Analytics Now, Post45, http://post45.research.yale.edu/2019/05/distant-reading-and-literary-knowledge/#identifier_5_10413.

6 *Merriam-Webster,* s.v. "theory," https://www.merriam-webster.com/dictionary/theory.

7 See Katherine Bode, *A World of Fiction: Digital Collections and the Future of Literary History* (Ann Arbor: University of Michigan Press, 2018); Tara McPherson, *Feminist in a Software Lab: Difference and Design* (Cambridge, Mass.: Harvard University Press, 2018); Whitney Trettien, *Cut/Copy/Paste: Fragments of History* (Minneapolis: University of Minnesota Press, 2019), https://manifold.umn.edu/projects/cut-copy-paste.

8. Catherine D'Ignazio and Lauren Klein, *Data Feminism* (Cambridge, Mass.: MIT Press, 2020).

9 Andrew Piper, *Enumerations: Data and Literary Studies* (Chicago: University of Chicago Press 2018), 11.

10 Michael Weisberg, "Modeling," in *The Oxford Handbook of Philosophical Methodology,* ed. Herman Cappelen, Tamar Szabó Gendler, and John Hawthorne, 262–86 (New York: Oxford University Press, 2016).

11 I am grateful to Jed Esty, who developed this point in his concluding remarks at the Imagined Forms conference in 2017.

12 I. A. Richards, *The Philosophy of Rhetoric* (Oxford: Oxford University Press, 1936).

13 Shakespeare's famous lines about the stage are also an example of what N. Katherine Hayles calls "material metaphors," symbolic moments in a text when the image or idea in the text reflects something of the material basis of the text (the medium). Hayles is interested in moments when the subject of our reading or viewing turns into the object of our reading and viewing. See Hayles, *Writing Machines* (Cambridge, Mass.: MIT Press, 2002).

14 Richard So, "All Models Are Wrong," *PMLA* 132, no. 3 (2017): 669.

15 So, 668.

16 Jennifer Egan, *Look at Me* (New York: Anchor Press, 2002), 40.

17 Egan, 179.

18 Egan, 181.

19 Egan, 181.

20 Ellen D. Wu, *The Color of Success: Asian Americans and the Origins of the Model Minority* (Princeton, N.J.: Princeton University Press, 2015).

21 It is impossible to name all the scholars who have undertaken such work. A brief, incomplete list includes postcolonial thinkers such as C. L. R. James, Étienne Balibar, Vivek Chibber, and Robert Young; Black Marxist scholars Angela Davis, Robin D. G. Kelley, and Cedric Robinson; Marxist feminists Silvia Federici, Nancy Fraser, Donna Haraway, Janice Radway, and Marion Young; and scholars of queer Marxism, including Kevin Floyd, Petrus Liu, and Christopher Nealon. Rey Chow has frequently critiqued the tendency to focus on immaterial labor by clarifying how Marx and Marxist thinkers shift between things and relations. See Chow, *The Protestant Ethnic and the Spirit of Capitalism* (New York: Columbia University Press, 2002). For a useful summary of efforts to theorize "concrete" labor in Marx, especially by Black, feminist, and Black-feminist scholars, see Michael Ekers and Alex Loftus, "On 'the Concrete': Labour, Difference, and Method," *Antipode* 52, no. 1 (2019): 78–100.

22 Combahee River Collective, "A Black feminist statement," 1977, *Monthly Review* 70, no. 8 (2019), https://monthlyreview.org/2019/01/01/a-black-feminist-statement/.

Acknowledgments

This volume was inspired by the Hagley Museum and Library's fantastic collection of patent models, which we, the coeditors, had the great fortune to visit (with special thanks to David Allen Cole and Erik Rau). Working together with Professors Jason Hill and Lance Winn, we posted a conference call to see if there might be some curiosity in studying models more broadly and addressing the materials, methods, and meanings of miniature representations. More than eighty responses from the widest range of fields—including art, architecture, comparative literature, digital humanities, English, history of science and medicine, media studies, music, and women's studies—convinced us that, yes, there was in fact more than a little interest in the topic. With the generous support of the Center for Material Culture Studies at the University of Delaware, as well as the Hagley, we brought together nearly two dozen speakers from across the country and punctuated two days of discussion with two keynote speakers, Johanna Drucker and Peter Galison. We thank them; we thank the conference speakers; we thank the many people who responded so enthusiastically to our call; and we thank the many people who attended the conference. We are grateful to Matthew Bird, Anne Bruder, Maura D'Amore, Eric Gollanek, Beverly Grindstaff, Zack Lischer-Katz, Yael Padan, Janelle Rebel, Margaret Simon, Anna Toledano, Shirley Wajda, Emily Warner, and Patricia Yu for their wonderful contributions at the conference and to Jed Esty for his artfully synthetic concluding remarks. Jessica Venturi and Michael Doss, the center's graduate assistants, helped with everything from managing the website to program designs to conference registration. Thanks to Heather Bohler and the Hagley team for being the perfect hosts. The University of Delaware's College of Arts and Science and its Departments of Art History, English, and History provided crucial support. Dean George Watson was a generous advocate for the conference. When it came time to support the volume financially, Kaylee Olney worked serious magic. Thanks to John Ernest and the Departments of English and Art History and to Lauren Petersen and John Pelesko in the Dean's Office for their generosity.

We are incredibly grateful to every one of this volume's contributors—without their intellectual curiosity and far-ranging insights, this volume would not exist.

The pleasure of working on a volume such as this one is learning from the ideas that the contributors have so generously shared. We are especially thankful that they stuck with us under the impossible conditions of Covid-19, finishing their essays and securing illustration permissions despite endless professional and personal challenges. The librarians and archivists who helped our contributors obtain their illustrations deserve special thanks.

How lucky we are that this volume has found its ideal home at the University of Minnesota Press! Many thanks to Pieter Martin for his unwavering support of the project, his patience, and his good humor throughout. We are also thankful to Doug Armato for sharing the project with Pieter in its early stages and to Anne Carter for all of her editorial support. Holly Monteith's meticulous editing brought the volume into shape, and Derek Gottlieb's wonderful indexing made it searchable. The volume benefited immensely from the comments offered by the readers for the Press, Steven Lubar and an anonymous reader; their rigorous feedback and queries on each individual essay and on the volume as a whole helped us to clarify and unify the project.

We would be remiss not to thank the staff of the Cherry Street Tavern, the best space for collaborative work in Philadelphia. Our work as editors on this project was supported in ways material and immaterial by our friends and family. Although we do not name them all here, their love, kindness, and ability to tolerate us when all we can think about is modeling is really what makes volumes like this possible. Coediting and coauthoring are agreements entered on the basis of little more than goodwill and hope. In the case of *Modelwork,* these arrangements crystallized into genuine partnerships characterized by openness, trust, mutual reliance, and abiding respect. Each of us thanks the other for clarifying each other's ideas, sharpening prose, and picking up the work when other matters intervened, as they always do. This project was in every way a model of collaboration.

Contributors

Martin Brückner is professor in the Department of English and director of the Winterthur Program in American Material Culture at the University of Delaware. He is author of *The Social Life of Maps in America, 1750–1860* and *The Geographic Revolution in Early America: Maps, Literacy, and National Identity*; editor of *Early American Cartographies*; and coeditor of *Elusive Archives: Material Culture in Formation* and *American Literary Geographies: Spatial Practice and Cultural Production, 1500–1900*.

Hilary Bryon is associate professor of architecture at Virginia Tech.

Johanna Drucker is Bernard and Martin Breslauer Professor of Bibliography and Distinguished Professor of Information Studies at UCLA. She is author of *Graphesis* and *Visualization and Interpretation*.

Seher Erdoğan Ford is assistant professor of architecture at Temple University.

Peter Galison is the Joseph Pellegrino University Professor at Harvard University. He is author of *How Experiments End*, *Image and Logic: A Material Culture of Microphysics*, and *Einstein's Clocks, Poincaré's Maps: Empires of Time* and coauthor of *Objectivity*.

Lisa Gitelman is professor of English and media studies at New York University. She is author of *Paper Knowledge: Toward a Media History of Documents*.

Reed Gochberg is lecturer in the history and literature program at Harvard University.

Catherine Newman Howe is lecturer in art history and studio art at Williams College.

Sandy Isenstadt is professor and chair of the Department of Art History at the University of Delaware. He is author of *Electric Light: An Architectural History* and *The Modern American House: Spaciousness and Middle-Class Identity.* He is coeditor of *Cities of Light: Two Centuries of Urban Illumination* and *Modernism and the Middle East: Architecture and Politics in the Twentieth Century.*

Christopher J. Lukasik is associate professor of English and American studies at Purdue University. He is author of *Discerning Characters: The Culture of Appearance in Early America.*

Martin Scherzinger is associate professor of media studies at New York University. He is editor of *Music in Contemporary Philosophy* and composer of *African Math.*

Juliet S. Sperling is assistant professor of art history and Kollar Endowed Chair in American Art at the University of Washington.

Sarah Wasserman is associate professor of English and associate director of the Center for Material Culture Studies at the University of Delaware. She is author of *The Death of Things: Ephemera and the American Novel* (Minnesota, 2020) and coeditor of *Cultures of Obsolescence: History, Materiality, and the Digital Age.*

Annabel Jane Wharton is the William B. Hamilton Professor of Art History in the Department of Art, Art History, and Visual Studies at Duke University. Her books include *Building the Cold War: Hilton International Hotels and Modern Architecture*; *Selling Jerusalem: Relics, Replicas, Theme Parks*; *Architectural Agents: The Delusional, Abusive, Addictive Lives of Buildings* (Minnesota, 2015); and *Models: Buildings, Bodies, Black Boxes.*

Index

Page numbers in italics refer to illustrations.